SAGE CONTEMPORARY SOCIAL SCIENCE ISSUES 34

CITIZEN

PREFERENCES

AND URBAN

PUBLIC POLICY

Models, Measures, Uses

Edited by

Terry Nichols Clark

 SAGE PUBLICATIONS *Beverly Hills / London* 1976

97379

The material in this publication originally appeared as a special issue of POLICY AND POLITICS (Volume 4, Number 4, June 1976). The Publisher would like to acknowledge the assistance of the special issue editor, Terry Nichols Clark, in making this edition possible.

For information address:

SAGE PUBLICATIONS, INC.
275 South Beverly Drive
Beverly Hills, California 90212

SAGE PUBLICATIONS LTD
St George's House / 44 Hatton Garden
London EC1N 8ER

Printed in the United States of America
International Standard Book Number 0-8039-0749-4
Library of Congress Catalog Card Number 76-53963

FIRST PRINTING

CONTENTS

Introduction
 TERRY NICHOLS CLARK 5

Modes of Collective Decision-Making: *Eight Criteria for
 Evaluation of Representatives, Referenda,
 Participation and Surveys*
 TERRY NICHOLS CLARK 13

**The Capabilities of Voting Rules in the
 Absence of Coalitions**
 T. NICOLAUS TIDEMAN 23

**Comments on Preference Revelation for
 Public Policy Decisions**
 JEROME ROTHENBERG 45

The Democratic Response of Urban Governments:
 An Empirical Test with Simple Spatial Models
 WAYNE HOFFMAN 51

Utility and Collectivity: *Some Suggestions on the
 Anatomy of Citizen Preferences*
 G. DAVID CURRY 75

Using Budget Pies to Reveal Preferences: *Validation of
 a Survey Instrument*
 JOHN P. McIVER
 ELINOR OSTROM 87

**Measures of Citizen Evaluation of Local
 Government Services**
 DOUGLAS SCOTT 111

Citizen Surveys for Local Governments: *A Copout,
 Manipulative Tool, or a Policy Guidance and
 Analysis Aid?*
 HARRY P. HATRY
 LOUIS H. BLAIR 129

Notes on Contributors 141

Introduction

Terry Nichols Clark

This volume emerges from three general sources: demands for enhanced citizen inputs to the political process; models of democratic processes; and survey research on public policy. Here, I offer a comment on each.[1]

DEMANDS FOR ENHANCED CITIZEN INPUTS TO THE POLITICAL PROCESS

The late 1960s saw dramatic increases on many political systems for enhanced citizen inputs. The urban disorders involving Blacks in many American cities and student and worker demonstrations in Europe received most publicity. In their wake have come numerous efforts to increase citizen inputs. Decentralization and increased citizen participation in neighborhood governments and places of work were frequent demands. To respond, various institutional structures along these lines were tried, ranging from citizen boards for Model Cities to little city halls. By the early 1970s, many public officials were convinced that efforts at increasing direct citizen participation had not been effective. Some turned to surveys of citizens as an alternative (indeed one accepted by certain American federal agencies as a necessary component for grant applications by local agencies). Evaluations of social programs also often include citizen surveys. What do social scientists have to contribute to this effort?

MODELS OF DEMOCRATIC PROCESSES

Without inordinate simplification, two general approaches to modeling democratic processes may be distinguished. The two have quite different implications for citizen inputs. The first goes under such names as group theory, pluralism, or the elitist theory of democracy. It emerges from such writers as A. Bentley, J. Schumpeter, V.O. Key, and R.A. Dahl. There are many variations, but most suggest that few individual citizens are sufficiently interested in specific public issues to become informed enough to take a clear position. Hence it is 'rational' for a citizen to support a political leader or a union or a party label that he has came to trust, and to leave specific issues and tactics to other actors. This approach cautions us to the demands which can be made on citizens concerning specific issues. Insofar as surveys of the general public are undertaken, this approach suggests they might deal with quite broad matters such as support for political candidates or parties. At the local level, one might interview leaders, or ask citizens about general support for programs like Model Cities, if not specific policies. If one does question citizens about specific policy issues, they should concern only those issues about which the citizen is reasonably certain to be well informed: personal experience with police officials or victimization by criminals, how often garbage collectors come by, and so forth.

The second general approach to modeling democratic processes is the populist or economic model of democracy. Recent advocates include A. Downs, J.M. Buchanan, and G. Tullock. These three and many others have sought to conceive of the political system as analogous to a pure market economy. Citizens then become the only sovereigns and it is their preferences which competing candidates for public office seek to implement. If candidates (political entrepreneurs) are successful, they will deliver public policies with just as invisible a hand as the pure market entrepreneur delivers private goods. Hence, intermediary groups such as political parties, pressure groups, unions, etc. are relegated to a largely subordinate role, certainly by contrast with the role they play in group theories.

The group theory approach has been the dominant tradition in American political analysis since the Second World War. But since the late 1960s, the populist approach has been gaining ground. If group theories were often less conceptually systematic, they were firmly grounded in empirical studies (of national interest groups, local decision-making, etc). By contrast, the populist theories have been highly abstract, often mathematically rigorous, but sadly lacking in empirical evidence. This has recently been changing.

SURVEY RESEARCH ON PUBLIC POLICY
Since the 1930s, survey research on political questions has mushroomed. But most surveys by Roper or Gallup, as well as social

scientists, have focused on support for political parties, individual candidates, or quite general issues, such as confidence in the president. There has been very little work addressed to specific public policies, such as should we increase national health insurance, or how satisfied are you with local police protection? This lack of work is not without cause. Probably the leading statement on these matters was long that of P.M. Converse[2] whose analysis of survey questions concerning public policies suggested that their fluctuations were so great — largely due to citizen information — that detailed study of specific policy issues was generally unfeasible in surveys administered to the general public. More recent work has suggested that citizen information and interest in public policies has increased since Converse and others reached their conclusions.[3] But just how far one can go, just how much the survey analyst may legitimately ask of the general public, remains very much an issue of debate. Much depends on the context and format of the questions, recent public discussion, and related situational issues. In the last few years, surveys of citizen preference, undertaken to help inform public policy, have grown dramatically. Most have simply ignored the Converse-type criticisms, and pushed ahead. For example, the State of North Carolina has recently undertaken a survey of citizen spending preferences using budget pies. This activity has grown sufficiently large in volume and policy import to demand assessment. Such is our purpose.

*   *   *

This volume grows out of a session on Preference Revelation for Public Goods, held at the Public Choice Society meeting in Chicago, 3-5 April 1975. Presented at the session were draft versions of the papers by Tideman, Hoffman, and Rothenberg, as well as two papers that appear elsewhere.[4] Comments by Hatry and Ostrom at the session later grew into their papers found here.

The Public Choice Society is a leading forum for the application of economic analyses to political phenomena. This partially accounts for the attention in this volume to such questions. Unlike much earlier work in the public choice tradition, however, the papers here show direct concern for assessing general models and concepts using systematic evidence. In the process, some fundamental questions are raised about basic assumptions and approaches in the 'economic theory of democracy'. Several differences of view distinguish the contributors. Nevertheless many of the results complement one another and point in similar directions.

It would be false modesty not to suggest that, taken together, these papers probably represent the most serious and detailed assessment to date of the uses and limitations of citizen surveys of urban public policy.

The papers fall into four groups. The first group, including those by Clark, Tideman, and Rothenberg, pose general questions about how to assess and measure citizen preferences for public goods.

Clark considers four 'modes' of decision-making — majority rule in selecting representatives, referenda, political participation and citizen surveys. The paper differs from other efforts at assessing such modes in the range of criteria applied: economic efficiency, determinateness, changes in information and tastes, preference structures, intensity of preferences, strategic behavior, decision-making costs, responsiveness to leadership qualities. Building on several years of experience of working with the demanding assumptions of abstract preference models, the paper reminds us of many apparently forgotten strengths of participation and representative government.

Tideman, like Clark, reviews several modes of collective decision-making, but the criteria he emphasizes are economic efficiency, equality, stability, and workability. The modes considered include majority rule, but the paper is especially valuable in its assessment of a wide range of additional modes of collective decision-making. Many of these are quite complex schemes never tried in practice, but which collective choice theorists have invented better to approximate (most often) economic efficiency and to minimize strategy. Many theoretical considerations that led to the surveys with budget pies in later papers are clearly presented here.

Rothenburg focuses on the little-discussed issue of using semi-homogeneous subgroups of citizens. He is critical of such a political cultural approach, instead arguing in favor of analyzing discrete individuals. His analysis leads to a quite different result from Curry's considerations of a similar problem.

The second section includes papers by Hoffman and Curry which test and extend median preference theories. Both analyze survey data about citizen demands for spending on a range of urban public policies. Both start from a neo-Downsean perspective; neither finds convincing support for this perspective.

Hoffman's central hypothesis is that if local political leaders follow a median preference strategy, then citizen preferences should be equally divided between those who prefer more and those who prefer less spending. Using data from ten American cities, he finds that citizens quite consistently prefer more services than they are receiving. But they also report that taxes are too high. These results hold even when several adjustments are added for income, race, political participation, etc.

Curry extends a neo-Downsean model to include benefits that individuals receive by virtue of group membership. He then analyzes citizen data for Boston to see how well demands for public services can be explained by membership in ethnic and other groups. Although several measures of group membership are statistically significant, the regression results show that a wide range of variables explain citizen policy preferences only very minimally. In particular, he finds that neither the Irish nor Blacks stand out, in contrast to earlier results of Clark and Hoffman which showed these two groups to be quite distinctive supporters of public expenditure.

The two most obvious interpretations for these results of Hoffman and Curry seem to be: (1) citizen preferences are much less structured than many neo-Downsean interpretations suggest; and/or (2) the specific survey formats were too crude to elicit information of the sort called for by the abstract preference theories.

The third group of papers, by McIver and Ostrom and Scott, speak directly to the survey format question. Both employed budget pie survey formats with large samples of citizens. Both included budget pies for their presumed theoretical superiority to simpler questionnaire formats. But both present results which raise serious questions about the budget pie format.

McIver and Ostrom summarize results from a survey of 1401 citizens in the St Louis metropolitan area which included a budget pie for three police services. They analyzed the data in several ingenious ways. Responses to an open-ended question asked what one improvement should be made with money from a federal or state grant for law enforcement. Resulting allocations hardly correlated with budget pie responses. Actual expenditure allocations by each citizen's police department were then compared with budget pie results — again correlations were minimal, although significant for higher SES neighborhoods. Next, seven items tapping evaluations of the police in terms of honesty, courtesy, etc. were correlated with the budget pie responses; correlations were modest, although often statistically significant. Finally, to assess strategic response patterns, four groups were examined, classified by jurisdiction size and level of information about the police. As hypothesized, citizens in larger jurisdictions and with more information seemed to respond more strategically.

Scott's work provides the most direct effort at methodological comparison of budget pies with other formats. He administered questions about urban public services (street repair, etc.) to 1028 citizens in the Los Angeles area. The same basic questions were repeated using four distinct formats: a budget pie, card sort, ordinal ranking, and self-anchoring scale. The last three methods generated fairly high (about 0.6) intercorrelations, and meaningful scales. But these three methods correlated insignificantly with the budget pie.

McIver and Ostrom report difficulty in using the budget pie format with lower status respondents; 21 per cent of their sample did not complete the budget pie properly, although only 4 per cent of Scott's respondents refused or had no opinion.

The last paper confronts several issues in the use of surveys for public policy. Although all the papers have suggestions about how and where different types of surveys are appropriate in policy making, Hatry and Blair comment directly from their experiences in helping local public officials use citizen surveys. Their judicious remarks about the uses and limits of surveys deserve serious attention. They emphasize the value of surveys for specific, realistic questions about which the citizen is knowledgeable and interested. They list several examples (and include more detail in related publications). Cost estimates are also presented. Unlike most other papers, they directly address the impacts that surveys may have on local officials and citizens. In particular, they consider it a 'cop out' for local officials to ask citizens about general policy priorities (as is done with most budget pies). The issues are simply so complex, they suggest, that 'an informed response requires a knowledge of the costs and benefits . . . which the average citizen has little likelihood of having'. Their remarks close the volume on a note of caution similar to that with which Clark opened it.

NOTES

1. This volume is part of a project generously supported by USPHS, NICHD, HDO8916-02.

2. P. Converse, 'The Nature of Belief Systems in Mass Publics', in D. Apter (ed.), Ideology and Discontent (New York: Free Press 1964) chapter 6.

3. See recent issues of the American Political Science Review and forthcoming books by S.Verba and N. Nie and B. Page.

4. T.N. Clark, 'Can You Cut a Budget Pie', Policy and Politics, Vol 3 (December 1974), 3-32; B. A. Scherr and E. M. Babb, 'Pricing Public Goods', Public Choice, Vol. 22 (Fall 1975), 35-48.

Modes of Collective Decision-Making: Eight Criteria for Evaluation of Representatives, Referenda, Participation and Surveys

Terry Nichols Clark

This short paper seeks more to raise questions than provide answers. It differs from most recent work on measuring citizen preferences in the range of questions posed. Indeed, most such work, we argue, represents what economists term suboptimizing. The major studies on social welfare functions deserve our respect for their mathematical elegance.[1] But how is this tradition to be linked to more general concerns, such as assessing legislative decision-making versus citizen participation versus surveys? This paper outlines a framework that might be applied in a general benefit-cost approach. If its answers, in the form of entries in Table 1, are only most tentative, it still points to a distinctive set of questions for future work. In the meantime, policy makers and their consultants must reach decisions regarding the relative merits of such things as citizen surveys. This framework may serve minimally as a structure for assessment of alternatives.

Table 1 lists four *decision-making modes* and eight *criteria for their evaluation*. The remainder of the paper comments briefly on the various cells of Table 1.

I. FOUR DECISION-MAKING MODES

The four modes in Table 1 are singled out because of their empirical frequency. Some discussion applies to these four in general, but most examples are from recent political decision-making in the United States. It would be suggestive to review these and other decision-making modes in quite different national contexts.

This is Research Report 48A of the Comparative Study of Community Decision-Making.

Table 1. Four Modes of Decision-making and Eight Criteria for Their Evaluation.

A. Majority Rule in Selecting Representatives	B. Majority Rule in Direct Referenda on Binary Choices	C. Political Participation	D. Citizen Surveys	
H	H	L	M	1. Economic Efficiency
M	M	M	M	2. Determinateness
M	H	M	H	3. Changes in Information and Tastes
H	M	M	M	4. Preference Structures: Homogeneity, Unidimensionality, etc.
L	L	H	D	5. Intensity of Preference
L	H	M	H	6. Strategic Behaviour
L	M	H	D	7. Decision-Making Costs
H	L	M	D	8. Responsiveness to Leadership Qualities

H = High; M = Medium; L = Low; and D = Depends on Specific Format.

A. *Majority Rule in Selecting Representatives.* City councillors and legislators have long been chosen to 'represent' geographically defined constituencies. Although they seldom articulated their remarks in such general terms, critics of democratic government in the late 1960s frequently attacked this decision-making mode. If 'the people' were contrasted with their representatives as more legitimate sources of decisions, empirically 'the people' were often interested minorities and their advocates.

B. *Majority Rule in Direct Referenda, Binary Choices.* In American states and cities, Swiss cantons, and French national decisions, the referendum has become common. With cable television, the technology permits instant referendum in the homes of all viewers. This spectre should encourage more thoughtful consideration of its effects than it has received to date. Limiting choices of binary (yes-no) decisions simplifies while restricting information elicited.

C. *Political Participation* encompasses acts ranging from letter writing to meetings of citizen groups to organized lobbying by interest groups. It is at once the most varied in form and the hardest to delimit in impact. It is alone indeterminate, and becomes a 'mode' of decision-making only by complementing other, more structured processes. Nevertheless, leading aspects of political participation can still be contrasted with the three other modes.

D. *Citizen Surveys* may take forms varying from the yes-no poll on a salient issue, to degree of 'confidence' in the current administration, to a choice among leading candidates for public office. More sophisticated questionnaire items may include Likert-type scales, card sorts into different categories, or budget pies for the respondent to cut into pieces.[2] Insofar as it differs from a referendum, the survey is less directly binding on its users. Although means, medians, or more complex summary statistics can be computed from survey data, these normally provide just one input for leaders to assess with others in reaching a decision.

II. EIGHT CRITERIA FOR EVALUATING DECISION-MAKING MODES

1. *Economic Efficiency* we list first as it has been at the center of much earlier work, and it remains the measuring standard for most economists. Economy efficiency may be broadly defined as an ideal 'equilibrium' situation where each individual's preferences are optimally implemented in all decisions which affect him. It is a social situation which cannot be altered without making some member of the collectivity worse off. The efficiency concept obviously derives from the perfect market analysis of private goods, where equilibrium occurs when the marginal rates of substitution among consumption of goods and services equal the marginal rates of substitution of production of goods and services. Problems arise, of course, when 'externalities' enter, such that costs and benefits for one individual spill over on to

others. Where externalities are so great that consumption of a policy is potentially equal for all members, we are in the realm of *pure public goods* (to use Samuelson's 1954 definition). Following Wicksell, Buchanan and Tullock have argued that with public goods, economic efficiency can be achieved only with unanimity rule. Ronald Coase has suggested that, operating within various legal constraints, individuals can still bargain among themselves to achieve economic efficiency by trading off support for different decisions. But Coase's theorem, like many such analyses, holds only under highly demanding assuptions — including zero transaction and bargaining costs.

Majority rule in selecting representatives, or in referenda, or efforts to assess a median position in citizen surveys are all subject to the Condorcet paradox: the results may change depending simply on the order in which alternatives are presented. And even if votes on a range of alternatives are taken, intransitive results are likely unless the citizen preference structure is quite restricted (as discussed below).

If the criterion of economic efficiency is virtually never met by any of our decision-making modes, A, B and D still are sensitive to the preferences of all citizens. C (political participation), may also deserve 'medium' rank in Table 1 if we consider that it broadly captures intensity of preferences.

2. *Determinateness* is the degree to which a specific decision is generated by the decision-making mode. Simple mention of the social welfare function tradition suggests the degree to which determinateness has not always been stressed. K. Arrow, P. A. Samuelson and others have demonstrated that no determinate ranking of private and collective activities is possible — a social welfare function meeting their classical economic criteria (particularly efficiency) cannot be established. If this nihilistic result is accepted on its own grounds, one may still consider decision-making modes in terms of their *relative* determinateness. A and B rank high on this criterion, political participation (C) scores low, while surveys (D) fall in the middle.

3. *Changes in Information and Tastes.* Classical economic analyses and related approaches assume essentially zero information costs — each person will keep shopping for the optimal good — and fixed tastes — he will not change his mind while shopping or afterward. G. Stigler has argued that information costs may be treated like other costs: individuals seek to maximize utility functions where one constraint is information; the classical economic framework, he argues,

thus remains adequate. H. Simon has proposed instead that below a certain threshold, individual behaviour is subject of such 'irrationality' as intransitive preference orderings, etc. but that if an individual grows dissatisfied enough, he will seek more information and alternative solutions to the existing state of affairs. He need not spend so much of his time and energy as to achieve an optimal decision — defined in terms of maximizing his individual preference ordering. Rather, as soon as he transcends the threshold of dissatisfaction, enough to achieve a 'satisfactory' decision as far as he is concerned, he will stop looking. He will thus have 'satisfied' rather than optimized.

A similar argument can be made for the assumption of fixed tastes, equally integral to classical economic analysis. We are often members of collectivities where decisions are made that are low in salience to us; in such instances the marginal cost to us of obtaining further information is relatively high compared to other ways in which we may spend our time or money, so we 'rationally' remain poorly informed. Indeed, with just a little more information we might change our preference ordering altogether, but once this assumption of fixed tastes and the corresponding concept of consumer sovereignty — that what the individual consumer chooses is correct for him — are questioned, the whole framework of classical economics fails to generate its appealing results.

These criticisms are nearly as old as classical economics, but their implications for private goods markets are somewhat different from those for political decisions. Indeed, one of the strongest arguments for representative government is that voters do not know their interests. At least every voter is unaware of every one of his interests every minute of every day. And when concepts like short-term and long-term interests are introduced, not to mention simply inadequate representation of minors, the unborn, the mentally feeble, and non-voters (the last is often 50 per cent in local elections), the advantages of less populist conceptions of government seem to grow. Insofar as representatives may be chosen on some criterion (party-affiliation is the traditional example in Downs and elsewhere) which is more salient, simple, and less costly than acquiring information about specific policies, then representative government is superior to any mode of decision-making which depends on assessment of specific policies, such as referenda or surveys. This is the basis of the trusteeship conception of legislative leadership, which, even these brief comments suggest, can be incorporated into the traditional arguments for populist democratic decision-making.

How 'informed', 'rational', and 'interested' the mean voter in the US or elsewhere is is in part a matter of adequate definition and careful empirical work. The meaning of many results of voting and survey research studies are currently being debated, as are historical changes in the American electorate. Where some perceive only ignorance and apathy, others see the invisible hand waving to them behind the same data. (See recent issues of the *American Political Science Review* as well as the papers in this volume.) Without seeking to resolve such issues, we can suggest that the less shared is information in a collectivity, and the more tastes are subject to change, the less adequate will be those modes of decision-making that depend on considerable information and fixed tastes — referenda and surveys. Further, the less salient the policy is to actors in the collectivity, the less information they will acquire, and the less they will care about the specific outcome. It logically follows too that the less salient the policy to citizens, the more likely that they will value determinateness and lower decision-making costs over economic efficiency. Political participation has the advantage that apathetic or unsure persons can move to issues where their views are more firm.

4. *Preference Structures: Homogeneity, Unidimensionality, etc.* How well different decision-making modes perform depends also on the structure of citizen preferences. Consider first *homogeneity* in its logical extreme. If all members of a collectivity agreed on an issue, an economically efficient and determinate decision would be simple. All decision-making modes would generate the same result. But as the homogeneity of opinion decreases, more persons will be dissatisfied with any specific decision. Nevertheless, attitudes on basic issues are seldom randomly distributed; they fall into broad patterns which can often be identified with basic population characteristics. Political parties, the mass media, voluntary organizations and labor unions take stands and thus help structure the beliefs of persons who trust them. The more citizen beliefs are colinear with such organizations, the better are citizens represented by political participation of these organizations. A second issue is *dimensionality* of citizen preferences. For a city council decision on repaving streets, for most citizens higher and lower expenditure levels are probably two separate points on the same basic dimension. But 'law and order' may tap quite different dimensions for different citizens. For some Americans in the early 1970s it was an anti-black codeword. For others it was adherence to due process. For still others, it implied

expansion of the police force. With three separate dimensions, outcomes can no longer be arrayed as points along a single continuum. But introduction of multidimensionality dramatically complicates analysis.

Symmetry and *single-peakedness* are two additional aspects of preference structures. Symmetry refers to a similar distribution of preferences as one moves in either direction from the median postion. Single-peakedness obtains when there is a single most valued position — in contrast, for example, to a bimodal distribution. If symmetry and single-peakedness do not obtain, decision-making modes like referenda and surveys can generate cyclical and intrasitive results.

Heterogeneity, multidimensionality, symmetry, single peakedness and intensity are just some of the leading aspects of preference structures that enormously complicate reaching determinate, not to mention economically efficient, decisions. These matters obviously do not disappear when representatives or political participation are the principal modes of decision-making. It is more than possible, however, despite recent advances in analytical techniques, that responsible representatives are better able to approximate even economic efficiency than referenda or surveys.

5. *Intensity of Preferences.* A major criticism of voting for representatives and referenda is that they do not take account of intensity of preferences. They may indirectly. Non-voting is clearly a rough indicator of preference intensity. Nevertheless, the range of alternatives presented to an entire electorate must be so narrow that intense minorities on specific issues may have no direct means of expression at the ballot box. Political participation is clearly an effective mode for expressing intense preferences. Surveys may incorporate intensity of preference if properly designed. Simple yes-no polls do not (except crudely in the 'don't knows'). Likert-type scales (for example, a five-point continuum from strongly agree to strongly disagree) are better in this regard. The budget pie is superior to the Likert-type scale in that it records not only the size of each pie slice, but imposes a budget constraint in that the total must sum to 100 per cent. But such complex survey procedures as the budget pie, while reporting more information, are more demanding in their assumptions about information, preference structure, etc.

6. *Stragetic Behavior* by citizens, or advocating one goal in order better to attain another, has been a major concern of economists. Their

working assumption has been that if strategic behavior can be effectively used, it will be. The more complex schemes discussed by Tideman seek to eliminate possibilities for strategic behavior. Strategy can be most troublesome for simple surveys — apparently illustrated in the papers by McIver and Ostrom and by Hoffman in this volume (where citizens reported a desired increase in most service areas but also a decrease in taxes). Still, certain recent studies suggest that strategy is less widespread than many economists have believed. More generally, rather than dismissing a decision-making mode altogether because it may be distorted by strategic behavior, it seems of greater practical import to assess conditions under which strategy may be minimized. Eight situations for limiting strategy are presented elsewhere, including introducing uncertainty concerning decision-rules, using voters as representatives, redefining the individual's social context, and recognition by citizens of their changing tastes and high information costs.[3]

The more traditional modes of decision-making (A and C) are probably superior to many surveys in this regard, not because they eliminate strategy, but because they have been dealing with varying degrees of strategy as long as they have existed.

7. *Decision-Making Costs.* At the level of the citizen, this was dealt with above in terms of information costs and intensity of preference. Considered not from the standpoint of individual issues and citizens, but in terms of the viability of a political system, decision-making costs deserve more attention than they have received in the literature. It simply is impractical to hold referenda or conduct surveys on every decision in a political system. Once stated, the point seems obvious enough. It implies more generally that some sort of representative mechanism, probably accompanied by political participation of very interested individuals, is far more economical than frequent referenda or surveys. Most decisions are simply too routine to interest most voters. To seek to construct a political system on the opposite premise is simply naive.

8. *Responsiveness to Leadership Qualities.* Even if leadership consisted solely of mechanically implementing citizen preferences, the present state of the art is such that experienced representatives could almost certainly surpass survey analysts even at this task. (This comment should be seen not as defeatist, but as a call for more and better surveys.) More important, however, leadership often consists of much

more: anticipating and shaping, as well as following, citizen preferences, and more generally acting on all manner of subjects about which few citizens have specific views.

The evidence is by no means satisfactory, but some of the best work on American voters suggests that 'personality' and 'character' are of greater import in determining votes for candidates than are their positions on issues. Voting for candidates on the basis of ethnic and religious background is (for some analysts) a related 'non-policy' oriented criterion. And ethnic and religious balancing of slates is far more important in many American localities than candidates' positions on issues.[4] It has seldom been argued, but perhaps it should be, that many citizens are better equipped to select representatives in terms of character and social background than they are by their explicit statements of positions on issues. Indeed this may be the best way to select representatives even in terms of their issue orientations, given the minimal amount of information emitted by candidates and acquired by voters. This may seem an heretical statement in terms of the current emphasis on 'issue-voting,' and it may be more true for the US in the past than for the US and other countries in the future. The facile criticism of this position — that it accepts 'false consciousness' which would decline in a more ideological, especially multi-party system — may be partially correct. But until more serious work is done to demonstrate that this in fact is the case, it does not seem unjustified to seek to generalize from the best empirical work available. Perhaps a bit of editorial licence will also be permitted on this matter, especially since this paper is followed by several others that develop quite different views. The reader may judge for himself how much support for different decision-making modes is contained in the remainder of the volume.

NOTES

1. Two general reviews and criticisms of this body of work are T. N. Clark, 'Can You Cut a Budget Pie', Policy and Politics, Vol. 3 (December 1974), 3-22 and the paper by Nicolaus Tideman in this volume. Tideman's effort is broad in its coverage of decision-making modes, but narrow in its criteria: primarily economic efficiency, equality, stability and workability. By contrast, we here consider only four decision-making modes, but lay down a broader set of criteria. Obviously these criteria could be applied to other decision-making modes, such as those discussed by Tideman. (These two papers contain numerous references not included here.)

2. Such variations in format are at the center of attention in Clark, op. cit. and the papers by J. P. McIver and E. Ostrom and D. Scott in this volume. Here we stress more the general characteristics of surveys.

3. Cf. Clark, op. cit.

4. A good deal of material supporting this interpretation is reviewed in recent unpublished works by Benjamin Page, Dept of Political Science, University of Chicago.

The Capabilites of Voting Rules in the Absence of Coalitions

T. Nicolaus Tideman

This paper discusses the features of a variety of rules that might be used to make group choices. The rules are evaluated under an assumption that coalitions have no impact, as might be true when the number of voters is large and secret ballots are used. The contributions of a number of writers are discussed in a framework that exposes new relationships among a variety of ideas.

INTRODUCTION

Some conditions that bear on the merit of a voting rule are as follows: Are all choices to be made between pairs of alternatives, or will some choices have to be made over three or more options? If there are multiple options, are they easily located in a Euclidean space? Is it essential that the voting rule involve relatively little calculation, or can it require extensive use of a computer? Is it necessary to treat all voters equally, or can the votes of some be weighted more heavily than those of others? Is there any readily identifiable basis for determining such weights? Could the voters be trusted to reveal intensities of preferences, to be used as weights, or would they behave erratically or strategically if asked to do so? Would it be permissible to offer voters a chance to have their votes weighted more heavily in return for some payment? If voters were confronted with such offers and behaved strategically, what degree of risk aversion would enter into their responses? Depending on the answers to these positive and normative questions, more and more sophisticated rules may be feasible.

I am indebted to Martin Bailey, Peter Bernholz, James Buchanan, and Gordon Tullock for comments on earlier drafts of this paper.

DEFINITIONS AND CRITERIA

As the introductory paragraph stated, this paper is not concerned with the possible impact of coalitions. A coalition is not simply a group of voters who find their views coincident on a particular issue. In that sense, coalitions always exist. A coalition is a group whose members all vote the same way despite a preference of some members for a different outcome. An expectation that the favor will be returned motivates such behavior.

The decision to exclude the effects of coalitions from the analysis does not imply a judgment that their influence is insignificant. They are surely not insignificant in committees and legislatures. But the purpose of this paper is to elucidate the considerations that would affect the choice of voting rules for large electorates, where significant coalitions are costly to organize, and where secret balloting makes vote exchanging agreements unenforceable.

The criterion that Rawls[1] suggested for defining justice can be used for evaluating voting rules. What would roughly equal men agree on as principles for evaluating rules, if they did not know the scope of applications that might later be found for such principles? This is not an operational standard, but it can serve as a mode for organizing discussion.

If a person were indifferent to risk and cared only about himself, he would want the rules that maximized the expectation of the aggregate value to all persons, since that would maximize the expected value to him. However, it seems more likely that a person would be averse to risk, and would care somewhat about others, so that he would want to trade some expected aggregate value for distributional considerations.

If a person did not know what his own position in the income distribution would be, he might have a preference for rules that generated greater equality of incomes for several reasons. He might expect that the marginal value of money to him would be greater if he were poor than if he were rich. This might be termed general risk aversion. Or he might feel that his envy of the rich would exceed his envy of the poor, so that extra taxation for the rich for the benefit of the poor would have a net positive expected value. The same preferences for equalization would occur if he expected his compassion for the poor to exceed his compassion for the rich. Mitigating against a commitment to total equality would be a concern that stronger measures to promote equality would lead to the dissipation of resources, reducing incomes for all persons.

A concern that resources be well managed would not prompt

measures to promote inequality for its own sake. Rather, stability of entitlements would be sought.[2] The objective of stability in the distribution of wealth might also be the result of a particular type of risk aversion. A person may desire that as of any future time, his income not be subject to later uncertainty, even that which would generate equality. Such a preference might be termed 'particular risk aversion', or a preference for stability.

In the analysis that follows, efficiency, equality and stability will all be treated as relevant to the evaluation of voting rules. While voting rules could be evaluated solely in terms of these ultimate goals, it will be convenient to also consider a proximate goal of workability. This encompasses the questions of whether the processes and parameters required by particular rules could reasonably be expected to be developable, and whether voters can reasonably be expected to behave in the ways that they must for the rules to achieve their objectives.

To evaluate the efficiency of decisions, some concept of the value of group choices to individuals is needed. But an operational definition in the revealed-preference mold customary to economics is not feasible. Unless a person determines the group choice by his behavior, his behavior does not reveal the value of a choice. It is precisely that that generates the problems of voting theory.

We see inefficiency in voting despite the lack of an operational definition of value in the voting context because our theory tells us it will be there. But the assignments of individual value that are required for the inefficiency conclusion are generated in the minds of analysts rather than by external observations. In line with this, the statements above value to individuals that are made in the analysis that follows should be interpreted as hypothetical cases and not as potentially verifiable statements.

Even when value to individuals is interpreted as purely hypothetical, there is a further definitional problem that can arise when one attempts to define the value of the same change to many persons. Any uncertainty regarding the contribution of one person can affect the value of the project to other persons, since they may have feelings about what he ought to contribute. If one seeks only to define the value to all persons of one plan with definite financing, this is no problem. But conclusions about the value of the same plan under alternative financing require information or assumptions about how each person is affected by changes in the contributions of all others. In the discussion later of rules with variable financing, no provision is made for taking account of such effects.

An important related point is that the whole discussion of voting rules in terms of their likely consequences presumes that it is the consequences rather than the rules that people care about. It could be that people want a particular rule regardless of the consequences, simply for its own sake. This whole analysis is relevant only if such procedural concern can be tempered by consideration of consequences.

A person's preferences for rules by which group decisions are made may very reasonably depend on his beliefs about the types of questions to which the rules will be applied. If there were to be no restrictions on the agenda, risk aversion might lead a person to desire that collective action be limited to cases where sentiment was unanimous, lest the group decide to take everything he had. Ideally, the agenda would be limited to those questions that the rules would settle in the same way as they would have been settled from behind a veil of ignorance. The more likely it is that the rules will be applied to inappropriate questions, the greater will be the attraction of rules that restrict the possibility of changes.

RULES THAT TREAT ALL VOTES ALIKE — DISCRETE OPTIONS
Majority Rule
When there are only two alternatives, when compensation is not feasible, and neither alternative can claim the privilege that might be accorded to the status quo, it is hard to quarrel with simple majority rule. This rule might be justified by an appeal to the principle of insufficient reason. If we have no reason for asserting that the value of having his preference followed is greater for one voter than for any other, we may as well treat all preferences as equally intense and presume that the greater value lies with the majority.

One formal justification for the inference that efficiency lies with the majority has been offered by Bowen,[3] who suggested that intensities might be viewed as coming from normal distributions, in which case the preference of the median voter, which controls the outcome for majority rule, is in the same direction as the aggregate of all preferences weighted by their intensities. The assumption of a normal distribution of intensities cannot be verified or refuted when there are only two candidates; it simply serves as a formal justification. However, when there are more than two candidates the assumption becomes refutable: cyclic majorities would not occur if preferences were distributed normally. The possibility of cyclic majorities makes the extension of majority rule to cases of three or more candidates difficult and controversial. The detailed rankings of candidates by voters must be considered in some way, but how?

The Borda Rule and Its Extension

One way of taking account of the detailed rankings of candidates by voters is point voting, invented by Borda.[4] Under point voting, a candidate receives one point for each candidate he is listed above on each ballot, and the candidate with the most points wins. Black[5] suggests that if this criterion is used, it should be modified to allow for the possibility of ties in any voter's ranking. The adjustment is made by subtracting a point from each candidate's score for each candidate that is ranked above him on each ballot.

Black[6] finds the Borda point voting criterion lacking because it permits departures from the 'dominance principle' suggested by Condorcet, which states that if there is a candidate who beats all others in pairwise comparisons, then that candidate should be the winner. Figure 1 shows an example of a situation in which the

Figure 1

20	12	20	20	19	9
A	A	B	B	C	C
B	C	A	C	A	B
C	B	C	A	B	A

	A	B	C
A	--	51	52
B	49	--	60
C	48	40	--

dominant candidate is not the winner under point voting. The numbers above the columns on the left show how many voters place the candidates in the indicated order, with the highest-ranked candidate at the top. Thus there are twelve voters who rank A first, C second, and B third. The results of pairwise comparisons of the candidates are shown by the matrix at the right. The number in each cell is the number of voters who prefer the candidate indicated by the row to the one indicated by the column. For example, the 51 voters who prefer A to B are those who reported the first, second, or fifth order of candidates. The adjusted Borda count for each candidate is his row sum minus his column sum in the matrix, yielding A:6; B:18; C: -24, so that B is the winner by this criterion even though A beats B as well as C.

Whether this should be regarded as an argument against point voting depends on whether the dominance principle should be followed. One argument for following the dominance principle is that it is necessary to achieve consistency. But consistency cannot be treated as all-important because majorities may be cyclical, as in the example shown in Figure 2. A beats B, B beats C, and C beats A. A group

Figure 2

20	12	16	24	19	9
A	A	B	B	C	C
B	C	A	C	A	B
C	B	C	A	B	A

	A	B	C
A	--	51	48
B	49	--	60
C	52	40	--

decision rule must be able to operate when majorities are cyclical, and therefore we must be prepared when we have more than two alternatives to be inconsistent with what we would choose if there were only two alternatives. If a group decision rule operates to choose B in the example above, that is inconsistent with the choice that would be made if only A and B were available. For any choice there is an inconsistency. Arrow[7] showed essentially that the price of consistency in this type of situation is dictatorship. That price is clearly excessive. We must give up consistency. We cannot regard the preferences of majorities over pairs of options as decisive for decisions among more than two options. The question that remains is what compromise of consistency is most attractive.

Another objection to the Borda rule is that it does not distinguish between the votes that a candidate receives over a weak candidate and over a strong one. One way to overcome this objection would be to score each candidate in terms of overall popularity, and then calculate each candidate's score as the sum of votes over others, with each vote weighted by the strength of the candiate surpassed. One very simple way of assigning a weight to each candidate would be by the fraction of possible votes received. (One would not want to do this with just two candidates, because then their weighted vote scores would always be equal!) Thus in the example of Figure 2, the strengths would be A: (0.51 + 0.48)/2 or 0.495; similiarly B: 0.545; C: 0.460. The weighted vote totals would be

$$A: \quad 0.51 \times 0.545 + 0.48 \times 0.460 = 0.499$$
$$B: \quad 0.49 \times 0.495 + 0.60 \times 0.460 = 0.519$$
$$C: \quad 0.52 \times 0.495 + 0.40 \times 0.545 = 0.475$$

so that B would be the winner.

The weighted vote totals provide numbers that could be used as revised weights, leading to new weighted vote totals, which could be used as further revised weights, and so on to convergence.[8]

Voters would not be motivated to state their true preferences in such systems. A voter would generally put the strong contenders he favored at the very top of his ballot, the strong contenders he opposed at the very bottom, and weak candidates in the middle, to maximize the effectiveness of his ballot.

The Hare System

An alternative to the Borda system of point voting is the Hare system of successive eliminations, in which the candidate eliminated in each round is the one with the fewest first place votes in that round. Following each round, the remaining candidates are moved up on all ballots to fill the gaps left by removing one candidate. One serious shortcoming of the Hare system is that it is possible that a compromise centrist candidate will be eliminated in the early rounds for a lack of first place votes, even though he was the second choice of all or almost all of the voters. Coombs[9] suggests that to avoid this problem, the candidate eliminated should be the one with the most last place votes instead of the one with the fewest first place votes. This procedure is particularly appropriate if the candidates are believed to lie in a one-dimensional space, with each voter having 'single peaked preferences' among them. That is, each voter finds the candidates less and less attractive as he moves in either direction from his first choice. In that case, only the end candidates will receive last place votes, and the process will select the middle candidate that beats all others in paired comparisons.

A Rule of Estimated Centrality

However, if the candidates span a space of more than one dimension, one can no longer take last place votes as an accurate indicator of relative peripheral location. If the goal is to select the candidate that is closest to the center of voting preferences, then a procedure that Good and I[10] developed is applicable. We hypothesize that candidates may vary in an arbitrarily large number of attributes for which voters have preferences. For the sake of simplicity and workability we impose the assumptions that the indifference surfaces of voters for attributes of candidates are spherical, and that the discrete distribution of ideal points of voters may be treated for purposes of estimation as a continuous multivariate normal distribution. In the model, each voter

will return a ballot that places the candidates in the order of the
distances of their locations in attribute space from the voter's ideal
point. These assumptions permit a statistical estimation of the order
of candidate positions from the mode of voters' preferences, which
order is then adopted as the social rank ordering of the candidates.
It should be mentioned that to discover the winner of an election by
this criterion requires a complex computer program that has been
written so far only for the case of three candidates.

RULES THAT TREAT ALL VOTES ALIKE — OPTIONS IN A CONTINUUM

The Plott Problem

The discussion above is concerned entirely with decisions over discrete
alternatives. When possible group choices are conceived as points in a
continuum of one or more dimensions, a different set of problems is
encountered. Plott[11] has shown that a stable equilibrium under majority
rule, for two or more dimensions, could only occur at a voter's ideal
point that had the characteristic that the interests (most preferred
directions of movement) of other voters at that point could be paired
off in exactly opposite directions. This would be highly coincidental.

One way of looking at the necessary coincidence is as follows: If
a group starts at a point that is no voter's ideal point and considers
moving in a direction that is not by coincidence tangent to any voter's
indifference curve, then all the voters who do not benefit from motion
in the proposed direction will benefit from motion in exactly the
opposite direction. If there is an odd number of voters, one faction
or the other must be a majority. Only under the very special conditions
specified by Plott will there be a point where the faction in favor of
one direction always exactly equals the faction in favor of the opposite
direction. The voter who finds the initial point ideal votes against
all changes.

Hinich et al.[12] showed that the 'discontinuous' nature of voting is
critical to the Plott theorem. When the probability that a voter will
vote for an option is assumed to be a continuous function of his utility
for it and for the alternative, with reasonable restrictions on curvature,
then there is a unique option whose expected plurality against all other
options is positive. This makes intransitive majorities much less
overwhelming than the Plott result might suggest.

Selection of the Median

One way of circumventing the lack of equilibrium of majority rule in multiple dimensions is to have a choice procedure that does not consider alternative motions endlessly, but rather moves directly from voters' stated preferences to a selected option. In view of the Plott theorem, the selected option would generally be one that a majority of voters would reject in favor of some alternative, but since any choice would have that characteristic, the criticism is not devastating.

One such simple rule is to have each voter name his first choice for the outcome, and select the option whose co-ordinates are the medians in all dimensions of the stated preferences. (If the number of voters were even, one would take the midpoints of the intervals between the two median preferences in each dimension.) If each voter's preferences are single peaked such a point is a stable equilibrium with respect to proposals to change co-orindates one at a time. No such proposal can gather majority support.

There would be some incentives for misstatements of preferences under this rule, even if each voter's preferences go downhill from a single peak. An example is shown in Figure 3. A voter whose optimum

Figure 3

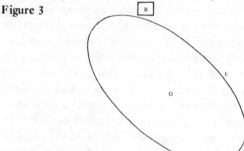

is 0, with an indifference curve given by the elipse E, would be able to influence the median one voter in each dimension, corresponding to the possibilities shown by the four corners of the box B. His first choice is the southwest corner, but the effect of stating his true preference is to have the southeast corner chosen, since that is closest to his optimum. If the voter could predict this outcome, he would move his stated optimum west, in order to have the effect he desired.

Minimizing the Sum of Distances

The multi-dimensional median has another feature that may be disturbing. It would be a different point if the co-ordinate system were rotated. A closely-related rule that would not have this limitation would be to select the point that minimized the sum of the distance from the choices of voters to the selection. This idea is discussed in Wendell and Thorson[13]. Note first that this rule is equivalent to majority rule in the case of a one-dimensional choice. Starting from a point that is not the median, the sum of the distance to all preferred points can be reduced by moving, say a distance d, in the direction of the median, since that reduces the distance to a majority of preferred points by d each, and increases by d the distances to a minority of points, so that the net effect is a reduction in total distance. The total distance is minimized at the median voter's choice if the number of voters is odd, and anywhere between the two median voters if the number of voters is even. These are the equilibria of uni-dimensional majority rule. The close connection of sum-of-the-distances with majority rule is further established by the fact that if voters indifference surfaces are concentric spheres and there is a Plott equilibrium, then the Plott equilibrium minimizes the sum of the distances to the voter's choices.

The rule of minimizing the sum of the distances from voters' stated ideal points to the point chosen may seem insignificantly different from taking the multidimensional mean, which is equivalent to minimizing the sum of the squares of the same distances. Actually there is a very crucial difference. If the announced choice procedure were to take the mean of the stated ideal points, any individual who thought he knew the direction from his ideal point to the point that would be chosen could bring the chosen point closer to his ideal point by simply stating an ideal point further from the point to be chosen, in the direction of his ideal point. The motive for greater misstatement would cease only when the voter thought he had moved the point to be chosen all the way to his ideal point. A system with a large number of voters engaged in such strategizing could not be expected to generate reasonable results.

On the other hand, if the criterion is the minimization of the sum of the distances, the ability of any one voter to influence the outcome is much more circumscribed. He cannot move the result outside the region where the absolute value of the gradient of the sum of distances, apart from the distance to his own ideal point, is less than or equal to one. This region is analogous to the box B in Figure 3. It is a convex shape with no ideal points of other voters in its interior.

Preference for the Status Quo
Throughout the discussion to this point, all options have been treated equally. However, if one option is the status quo, voters might prefer at a constitutional level that that option be given preferential treatment.

The preferential treatment of the status quo serves as a crude substitute for the right to compensation when one's expectations are not fulfilled. It would be more desirable to provide compensation for those who lose when a change is enacted, but if the difficulty of estimating the magnitude of individual losses makes it impossible to provide compensation as a practical matter, then people may reasonably prefer that a change be made only when the strength of the evidence that the change is an improvement is significantly better than fifty-fifty. The requirement of two-thirds majorities, in contests between two options, is one example of special treatment for the status quo. Other voting rules are amenable to corresponding special treatment for the status quo.

Weighted Voting

The rules discussed to this point treat all votes alike. Such equal treatment is a fundamental component of fairness when the issues under consideration are of a general nature, so that there is no reason to believe that any one voter's stake in the matter is different from any other voter's. However, it is quite possible that people would agree at a constitutional level that in particular issues the preferences of some persons should be weighted more heavily than the preferences of others. Weights proportional to the intensities of preferences, if they could be achieved, would guarantee selection of the most highly valued option.

From one perspective all the possible inefficiencies of voting may be viewed as arising from the fact that a 'yes' vote exactly cancels a 'no' vote. This practice promotes efficiency only if the intensity of the feeling behind a 'yes' vote typically equals the intensity behind a 'no' vote. We cannot observe the intensities, but we may have reasonable and shared beliefs that they are equal or not equal. One basis in particular for such a belief is predictability of vote patterns among groups.

Weights from Abstention Rates

If distinctly different voting behavior is observed among identifiable subgroups in the electorate, then a normal distribution of intensities of preferences, as required for the efficiency of majority rule, is relatively implausible. Still, if one could assume that within each subgroup the distribution of intensities was normal, and that an individual votes if, and only if, the intensity of his preference is greater than some critical minimum value, then the average intensities of preferences in subgroups can be compared.[14] The voters can then be arrayed in a distribution like Figure 4. An intensity of 0 falls midway in the abstaining range. The mean intensity can then be measured in multiples

Figure 4

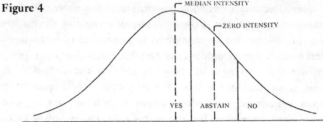

of the critical intensity required to provoke voting. If one accepts this interpretation of abstentions, the procedure could be used to estimate how often majority rule fails to produce efficient choices, but it would not itself be a better choice procedure.

Even apart from questions of how the subgroups would be defined, and whether the procedure could be considered fair, there is the fundamental problem that if a voter believed that the opposition within his subgroup would receive more than 50 per cent of the vote, then it would be in his interest to abstain rather than vote, because that would reduce the estimated average intensity by more. The reason for this is that if votes in the minority are changed to abstentions, the estimates of the number of standard deviations in the mean intensity and in the critical intensity needed to induce voting are increased by equal amounts. But since the former is greater when the majority has more than 50 per cent of the vote, their ratio, which is the estimate of the mean intensity, is diminished. Thus one could not design an efficient choice rule around this approach.

Weights Related to Tax Shares

If the voting patterns of subgroups do not vary randomly from one vote to the next but rather display some systematic differences, then there are other steps that might be taken to avoid the inefficiency of majority rule, particularly when the choices to be made involve expenditures. The formula for raising taxes to pay for an activity could be modified in the direction of raising the shares of those groups that tend to have a greater-than-average rate of approval, and lowering the taxes of the groups whose approval rates are less than average. If the analysis is accurate enough, the result will be that all groups will have nearly equal rates of approval; the suspicion that majority rule may be inefficient will no longer be supported by evidence of different group voting patterns.

When the tax shares of individuals for some activity have been set equal to their estimated benefits, there is a rationale for weighting the votes by the tax shares to increase the efficiency of the voting process. The rationale involves an assumption that the standard deviation of benefits is proportional to estimated benefits, and hence to tax shares. If this proportionality does exist, then the best estimates of the relative aggregate benefit in any group will be approximately the plurality (the excess of the majority over the minority) multiplied by the tax per voter. Thus, when the tax allocation is imperfect and only some groups have majorities in favour of a proposal, efficiency in the choice rule is probably promoted by weighting votes by taxes. The practice of having taxes proportional to benefits might be advocated not only to improve the efficiency of decisions but also to reduce the extent to which expenditure decisions generate random redistribution of income.

There are several significant limitations to the practice of setting tax shares and voting weights proportional to estimated benefits. The first is that it has no applicability to measures designed to affect the degree of income equality in a society. If a particular provision for redistribution is appropriate in the sense that people would agree on it when operating in ignorance of their actual positions, then the fact that people who know they are poor find the inequality excessive, while people who know they are rich see too much equality, is of no consequence. One cannot expect good redistributive changes to be generated by democratic processes. But this does not imply that it would be appropriate to institute rules of strict distributional neutrality in all allocative functions. Under the status quo, part of each person's real income is the public services that he can expect to receive by virtue of his franchise, exercised either directly on choices or through impact or representatives. Buchanan[15] develops this idea. Before one could move to allocative decisions made strictly on allocative merits without a distributional impact of that move itself, some settlement would have to be made of the claim that voters might make for the current distributional value of their voting rights.

Even if the distributional problem, which would apply to any method of rationalizing allocative decisions, could be handled, there would be a further problem of applying the vote-weighting rules fairly. Voters could potentially be classified in a myriad of ways, each of which would lead to a slightly different estimate of each voter's benefits from government activity. An 'econometric constitution' would be needed to specify the procedure by which significant variables would be identified and alternative functional forms compared. Furthermore, if any sensitive characteristic, such as race or sex, is forbidden as an element of classification, there will be a potential for uncorrected bias in allocation.

The entire discussion of weighted voting to this point has dealt with decisions between two alternatives. The same opportunities exist for weighting votes if the decision is among multiple alternatives, or for a continuous parameter. If a continuous expenditure parameter is involved, there will be an opportunity to revise cost shares, and hence weights, using the rule that any group that favours greater expenditure should be assigned a higher share. If the alternatives are candidates rather than expenditures, this opportunity to revise weights will not arise.

Weights from Surveys

From one perspective, the whole difficulty of assigning weights to votes arises from an unwillingness to trust people. Why not simply inquire of each voter the value to him of proposed projects, and carry out only those projects for which the net value was positive?

Approaches along this line have generally been disfavored by economists, because of the motivation for dishonest statements of preferences. However, Bohm[16] has offered experimental evidence that people may generally choose to ignore their strategic opportunities. Random samples of the residents of Stockholm were given Kr50 (about $10) and asked to value the opportunity to watch a particular TV program, having been told that the program would be shown if the sum of the stated values from the persons in the room, when added to the stated values of persons supposed to be elsewhere in the building, exceeded Kr500. Different samples were given different rules about how the fee for watching the program would be determined.

The rules varied considerably in the incentives they offered to overstate preferences, but the average preferences reported were not significantly different. This does not imply, however, that the means are truly the same, as the sample sizes were rather small. (Bohm says only that further research is warranted.)

The question posed by Bohm's work is whether a more extensive series of experiments of the same sort, if the resulting distributions appeared to be insignificantly different, should convince a reasonable person that it would be appropriate to rely on responses made under conditions providing an obvious selfish incentive for misstatements, to determine the average value of public activities. For me the answer is no. The magnitude of money and the psychological forces are so different in experimental situations than they would be under a real evaluation of public policy that relied on unpriced statements of individual value, that I do not believe I could be convinced that honest statements would be given in the evaluation of real policies.[17]

A further reason for not trusting procedures that are open to obvious distortion is that the potential gain in efficiency is limited to the difference between the median preferences, which results from methods derived from majority rule, and the mean preference, which would be most efficient. Whenever the distribution is symmetric, the mean and the median are identical. Any indication of a distinctly assymmetric distribution may be taken as evidence of a serious deficiency in the method of allocating the costs of public services. If the mean preference is greater than the median, then some group is being charged much less than the value to them. A mean less than the median is an indication that some group is being charged much more than the value to them. If the charges could be brought closer into line with the benefits, the mean and the median preference would tend to converge.

VOTING AS A GAME -- OPTIONS IN A CONTINUUM

It may be that preferences are markedly assymmetric, and no readily identifiable characteristic of voters will predict the strengths of preferences. If that is the case, and if people cannot be trusted to reveal the true intensities of their preferences when simply asked, we might still achieve efficient levels of public activities if the process of reporting preferences could be constructed as a game in which the strategy of reporting one's true preferences had the greatest expected payoff. The basic idea of efforts to construct such games is that people would be asked to report the value to themselves of marginal changes in the levels of public activities, and the reported values would be construed as offers to pay for services at that price, or alternatively as offers to accept reductions in public services provided taxes are reduced according to the claimed loss of value. A person who reports an artificially low value risks having his taxes reduced only slightly when services are reduced.

Entitlement to the Status Quo with Continuous Adjustment

Dreze and de la Vallee Poussin[18] suggested such a system in which individuals would report such marginal value parameters continuously, and the levels of public activities would be varied continuously. When the sum of marginal values exceeded marginal cost, the level of activity would be increased. When the sum of marginal values was less than marginal cost, the level of the activity would be reduced. The fundamental problem with this approach, of which Dreze and de la Vallee Poussin were aware, is that once the parameter starts to move, participants, knowing the direction of motion, will be able to determine what types of misstatements of their preferences will maximize their selfish interests. If an activity is being decreased, a claim of a very high value for that activity will generate a large decline in one's taxes. When a large number of people try to take advantage of such an opportunity at the same time, the rate at which the level of activity can change will approach zero, and all of the value of increased efficiency will accrue to those who are overstating the value of the activity. If the level of activity were seen to increase, strategists could claim a zero value. Again, the rate of change would slow to a virtual standstill, with all the gains captured by the strategists. A system that provides such an opportunity to use information about the direction of change in the level of an activity to obtain a selfish benefit is probably not acceptable.

Entitlement to the Status Quo with Discrete Adjustment

In 1972 I devised a related system in which strategic opportunities would be less pronounced.[19] The principal difference between the system I proposed and that of Dreze and de la Vallee Poussin is that adjustment in the levels of public activities would be discrete rather than continuous, so that voters could not be sure of the direction of change when they reported their marginal values of public activities. If they understated their benefits, their taxes would decline by less than their benefits in the event that the level of the activity was reduced. If they overstated their benefits, their taxes would rise by more than their benefits in the event of a rise in the level of the activity. A person who thought he could guess the direction of change in the level would have an incentive, if he were a gambler, for misstatements as in the system of Dreze and de la Vallee Poussin. But if he believed that the initial level was an unbiased estimate of the later level, then he could do no better than to state the true value of the marginal benefits to him.

Entitlement to the Consequence of One's Abstention

Clarke[20] offered a scheme for choosing the levels of public activities that provides incentives for honest statements even when the statements of other voters (and hence the direction of change) are known. The crucial difference between the two systems discussed previously and Clarke's is that the former give each voter an entitlement to the status quo: each voter is compensated for diminutions in public activity, and is required to pay for increments, only according to his stated evaluations. In Clarke's scheme, on the other hand, the voter's entitlement is to the outcome that would prevail if he were to abstain. This means that the collectivity can decide to produce additional services and charge him for his assigned share of the cost even if he claims that he receives no benefits. This alternate entitlement makes it possible to escape incentives for misstatements. The opportunities confronting a single voter in the Clarke voting system are shown in Figure 5.

Figure 5

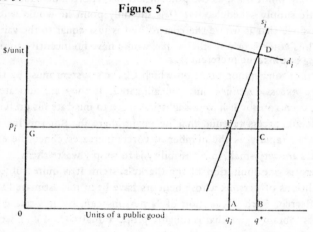

Without loss of generality, the marginal cost of the good may be taken as $1. (A dollar's worth is the unit of analysis.) Schedule s_i is the 'synthetic supply schedule' for voter i, the marginal cost schedule of the good ($1) less the contributions that have been offered by other voters. Schedule d_i is the true marginal value schedule of voter i. The line $\bar{p}_i$ is the cost share that has been assigned to voter i. The group decision rule proposed by Clarke is that the quantity produced be that where the sum of the offered cost shares is just equal to the marginal cost of production. If voter i reports his true demand schedule, this will be quantity q^*. The tax that would be assigned to voter i is expressed by Clarke as the voter's assigned tax share for that quantity, namely $\bar{q}_i$, that would be chosen if the voter reported his assigned tax share as his demand schedule, plus the integral under the voter's synthetic supply schedule from $\bar{q}_i$ to q^*. In Figure 5 this is rectangle OAFG plus trapezoid ABDF. If the voter's demand schedule were below $\bar{p}_i$ at q^*, his assigned quantity would be greater than q^*, and the second component of his taxes would be negative. In either case, the tax may also be expressed as his assigned price for the chosen quantity, plus a triangular figure bounded by price $\bar{p}_i$, quantity q^*, and the synthetic supply schedule s_i. This is rectangle OBCG plus triangle FCD in Figure 5. The sum of these rectangles for all voters is exactly enough to pay the cost of the goods to be provided, so Clarke's system generates excess revenue equal to the triangles' sum.

Suppose the voter whose situation is depicted in Figure 5 knew exactly how others had voted when he cast his ballot. He could then choose any point on s_i as the point where his reported demand cut the synthetic supply schedule. But D is the only point he would want to choose since that is where the cost to him is just equal to the value to him. Thus voter i, or any other voter, would have no incentive to reveal anything but his true preferences.

There is one minor point on which Clarke's system must be tidied up: the excess revenues must be allocated. If they are allocated to voters, each voter will have a slight incentive to misstate his preference to generate excess revenue that he might share in. But this incentive diminishes rapidly as the number of voters increases. Since the excess revenues are very small, one possibility is to simply waste them.

There is one limitation of the Clarke system. It is quite vulnerable to coalitions of persons whose benefits have been mis-estimated to the same degree. Such a coalition of N persons can, by misstating their benefits, obtain a gain that is roughly proportional to $(N-1)^2$. Therefore it would be very important, if Clarke's system were to be operated, that no person be able to know how any other person had voted, so that coalition agreements could not be enforced.

VOTING AS A GAME — DISCRETE OPTIONS
Entitlements Based on Assigned Probabilities
Earl Thompson[21] suggested a voting procedure for discrete options that would relate payoffs for outcomes to the prior probabilities of their occurrence. A bureaucracy would announce probabilities, and then voters would be permitted to buy insurance against their less-favored options, at prices determined by the announced probabilities. The option with the least insurance bought against it would be chosen, and the premiums collected on the other options would be used to pay off the policies on the chosen option. If voters agree with the

announced probabilities and are somewhat risk averse, they will maximize their utility by stating their preferences honestly.

The problem with this approach is that the rational behavior of a participant depends on his belief about the probabilities. There is no way to guarantee that all participants will have the same beliefs. For each individual there are announced probabilities that would motivate honest statements of preference, but the pattern that is right for one participant would induce gambling behavior in another. Still, there would be some prior probabilities and associated payoff pattern for which misstatements motivated by gambling would cancel, so that the choice would be the same if everyone had stated his honest preference. But it is not enough to suggest, as Thompson does, that the setting of the probabilities is a detail for bureaucrats to handle.

Every time a potential change is put on the ballot, any participant who favors the change and votes honestly increases his utility in dollars by the fraction of the value of the change equal to its assigned probability. Those who disfavor the change and vote honestly experience the corresponding loss. Such a system could be expected to work only if the assigned probabilities had some objective basis, since otherwise the process has a distributive impact that, to the individual participant, is random if not malicious.

Entitlement to the Consequence of One's Abstention

Clarke[22] devised a variant of his system for choosing among discrete options. It again requires each voter to pay according to his impact on the outcome. Each voter is asked to report the relative value of each option. The one with the highest aggregate value is selected. Any voter whose vote makes no difference in the outcome is charged nothing. A voter whose vote does make a difference is charged the difference between the aggregate value without his vote of the option that would win without his vote, and the aggregate value without his vote of the actual winner. As in Clarke's earlier system, no voter can do better than he can do by revealing his true preferences. But a coalition can hope to obtain their first choice without payment, even though it is not most efficient, by all stating excessive values, so that the subtraction of any one vote does not alter the outcome.

CONCLUSION

By way of conclusion, return to the factors mentioned in the introduction. How do they bear on the choice of a voting rule?

If choices are to be made only between pairs of options, and if weighting of votes is practically or morally impossible, then the best one can do is to rely on majority rule. It produces efficient choices when the intensities of preferences are distributed symmetrically. When there is some basis for estimating the intensity with which voters hold their preferences, weighting votes by such factors will improve the chances that majority rule will select efficient outcomes, and the weights can also be used as financing shares to cut down on distributional instability.

If votes are not weighted perfectly, then majority rule may select inefficient outcomes. If there are more than two possibilities, the errors of majority rule may be reflected in intransitivity of sequential majorities. A variety of devices are available for cutting through such intransitivity; Good and I propose that it be done by an indirect statistical estimation of the rank order of the distances of candidates from the center of voter preferences.

If choices are to be made about continuous parameters rather than discrete options, one can simply select the median preference in each dimension. A closely related rule that is independent of rotations of the co-ordinate system is to minimize the sum of the distance to the stated ideal points of voters. These rules would achieve efficiency only by coincidence, because votes are not weighted by intensity. Weighting votes by tax shares is sensible if the discrepancies between taxes and benefits are correlated with the level of taxes. Weighting votes according to stated intensities is sensible only if one wishes to be perfectly trusting.

Dreze and de la Vallee Poussin pointed out that for decisions about continuous variables, any reports of individual values that do not add up to cost can be used to move the levels of activities, with compensation, so that everyone's position is improved. However, once the direction of movement is established, strategic misstatements could be expected to slow the rate of change to a glacial pace. I propose a variation of this system in which an individual would only be allowed to set a single parameter of the benefit schedule attributed to him. This will motivate honest responses provided that individuals believe that the initial levels of activities are unbiased estimates of their final levels and provided that people are not risk preferers. Clarke proposed a system that motivates people to make honest statements about their schedules of marginal value even when they know how others will vote. The critical features of Clarke's system are prior assignment of cost shares, an entitlement to the consequence of one's abstention rather than to the status quo, and an entitlement of each individual to move along the offer schedule determined by other voters.

For discrete alternatives, Thompson proposed a voting game based on entitlements determined by assigned probabilities. There would be strategic responses if people did not all have the same prior probability, and it would be very difficult to get acceptable determinations of prior probabilities since their distributive impact can be so great. A version of Clarke's voting system provides incentives for honest statements of preferences about discrete options. A high degree of vulnerability to coalitions is the only serious limitation of such systems.

NOTES

1. J. Rawls, 'Justice as Fairness', Philosophical Review, Vol. 67 (1958), 164-94.

2. A comprehensive study of the role of and rules for income stability can be found in Frank Michelman, 'Property Utility and Fairness: Comments on the Ethical Foundations of "Just Compensation" Law', Harvard Law Review, Vol. 80 (April 1967), 1165-1258.

3. H.R. Bowen, 'The Interpretation of Voting in the Allocation of Economic Resources', Quarterly Journal of Economics, Vol. 58 (November 1943), 27-48.

4. For a historical discussion, see Duncan Black, The Theory of Committees and Elections (Cambridge: Univeristy Press 1963), 59-66, 156-59.

5. Ibid., 61-64.

6. Ibid., 57-59, 64-65, 159-78.

7. K. Arrow, Social Choice and Individual Values (New Haven:Yale University Press 1953).

8. For a discussion of such a system in terms of matrix algebra, see J.W. Moon and N.J. Pullman, 'On Generalized Tournament Matrices', SIAM Review, Vol. 12 (July 1970), 389-94.

9. C.H. Coombs, A Theory of Data (New York: Wiley 1964), 398-99.

10. I.J. Good and T.N. Tideman, 'From Individual to Collective Ordering through Multidimensional Attribute Space', Proceedings of the Royal Society (forthcoming).

11. C. Plott, 'A Notion of Equilibrium and its Possibility under Majority Rule', American Economic Review, Vol. 57 (September 1967), 787-806.

12. M. Hinich, J. Ledyard and P. Ordeshook, 'A Theory of Electoral

Equilibrium: A Spatial Analysis based on the Theory of Games', Journal of Politics, Vol. 35 (February 1973), 155-93.

13. R.E. Wendell and S.J. Thorson, 'A Mathematical Study of Decision in a Dictatorship' (The George Washington University Nonlinear Programming Symposium 1973).

14. This idea was suggested to me be Eli Noam.

15. J.M. Buchanan, 'The Political Economy of Franchise in the Welfare State', in R. Selden (ed.), Essays in Capitalism and Freedom (Charlottesville: University Press of Virginia 1975).

16. P. Bohm, 'Estimating Demands for Public Goods: An Experiment', European Economic Review, Vol. 3 (1972), 111-30.

17. At the Public Choice Conference in Chicago where this paper was presented, two other papers dealt with this point. B.A. Scherr and E.M. Babb, 'Pricing Public Goods', Public Choice, Vol. 22 (Fall 1975) discussed another experiment, analogous to that of Bohm, which seemed to uncover little strategic behavior. T.N. Clark, 'Can You Cut a Budget Pie?', Policy and Politics, Vol. 3 (December 1974), 3-31, suggested several procedures to minimize strategic behavior concerning public goods. The text has not been revised to deal with these arguments as my basic position remains unchanged.

18. T.H. Dreze and D. de la Vallee Poussin, 'A Tatonnement Process for Public Goods', Review of Economic Studies, Vol. 38 (April 1971), 133-50.

19. T.N. Tideman, 'Efficient Provision of Public Goods', in S. Mushkin (ed.), Public Prices for Public Products (Washington: The Urban Institute 1972), 111-23.

20. E.H. Clarke, 'Multipart Pricing of Public Goods', Public Choice, Vol. 11 (Fall 1971), 17-33.

21. E. Thompson, 'A Pareto-optimal Group Decision Process', Papers on Non-market Decision Making, Vol. 1 (1965), 133-40.

22. E.H. Clarke, 'Multipart Pricing of Public Goods – An Example', in S. Mushkin (ed.), Public Prices for Public Products (Washington: The Urban Institute 1972), 125-30.

Comments on Preference Revelation for Public Policy Decisions

Jerome Rothenberg

1. These are good papers. I found them rewarding. I shall have some remarks for all three, but most of this discussion refers to the Tideman and Clark papers (not, however, for reasons of quality). This Tideman paper is an exceptionally full survey, and very well done. There is solid analysis and reasonable judgement. It is quite a useful work. The Clark paper is most helpful to the reader in providing a great deal of background very clearly. The situational distinctions among the dimensions of public policy and citizen attitudes are especially illuminating.

Both papers explicitly specify criteria for evaluating different public decision rules. In view of the fact that these are often only implicit in public policy analysis, a potentially highly misleading procedure when there is such a variety of rules to examine, this overtness is an important strength.

Besides these rather similar general attractions, the Clark paper emphasizes that 'measuring instruments . . . (should be) . . . only as precise and complex as necessary for the decision at hand'. Since precision and complexity entail sometimes considerable cost to investigators or policy officials, this inherently economic principle is hard to resist. But this 'cautionary conclusion' deserves a cautionary remark: it is not always clear what degree of precision or complexity *is* needed for the decision at hand. The complexity of a situation depends on the perceptions and motives of the *participants,* not the observers. It is the behavior of the former that must be tolerably modeled before an observer can feel content that the situation has been

These comments refer to the papers presented at the Public Choice session in Chicago, as indicated in the Introduction to this issue [editor's note].

comprehended (and graded).

Scherr and Babb's, and Hoffman's experiments are quite relevant to this issue. They are posing the question: To what extent are certain complexities, alternatives of choice and strategies which are appreciated by observers of a particular class of situations actually present in the behavior of the participants in those situations? The conclusions of Scherr and Babb's paper suggests that there may be important discrepancies between these separate perceptions. Thus, moderating the demands of measurement to the demandingness of the situation may be itself a demanding challenge.

2. The surveys and the experiments are difficult to discuss in detail. There is neither the space nor the scope. But I shall make a few detailed comments and then a broader one.

3. On the measurement of preference intensities. This refers to three papers, in some degree.

(a) Contrary to the impression given the reader, especially in the Tideman paper, rank order forms of vote weighting, of the Borda and other types, do approximate the inclusion of preference intensity considerations into measurement, with comparisons of both intra- and inter-personal intensities. So, for example, if individual i prefers A to both B and Z, but A and B are effectively adjacent in i's preference ordering (i.e. approximately the same Borda score) while A and Z are widely removed in the preference ordering (very different Borda scores), then we may infer that A preferred to Z represents a greater preference intensity for i than A preferred to B. Such an inference is, of course, much more definitive than the comparison of i's preference for A over B with that of i for C over D when Borda score differences for the two pairs are equal. Its even less persuasive when comparing i's preference for A over B with individual j's preference for A over B, despite equal Borda score differences for each pair. Preference intensity inferences from rank order considerations have differing degrees of credibility but in some situations even interpersonal comparisons may be irresistible, and important to make.

4. Procedures designed to measure the mean preference intensity for whole groups, so that their preference directions can be given relative weights through these mean intensities, are even more questionable than Tideman's treatment suggests. The procedures attempt to infer group preference intensities from observed vote distributions via the assumption that individual preference intensities within the group are generated from normal distributions. They appear to demonstrate some misunderstanding of the relationship between preferences and preference intensities in the given contexts of choice.

(a) First, a distinction must be drawn between a normal distribution of preference intensities directly (independent of preference positions) and a normal distribution of individual preference positions on a policy alternative dimension. Since preference intensity seems to be a matter of a degree of difference in attitude toward more than one alternative rather than some intrinsic level of attitude toward a single alternative, inferences about intensities must be based on preference positions, which refer to differential attitudes toward different alternatives, and not supposed intensities for single alternatives. We shall therefore examine the inferential structure of these approaches under this former interpretation.

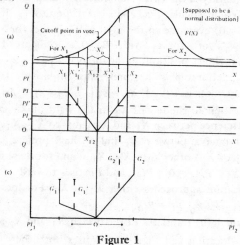

Figure 1

In Figures 1 (a) and (b), X is the identity of the policy alternative. The alternatives are assumed to be arrayable for all voters along a linear axis in such a way that distance along the axis is a representation of degree of preference difference.* PI represents amount of preference intensity in Figures 1 (b) and (c). Q represents the number of individuals who possess the characteristic indicated.

Assume an actual vote on alternatives X_1 and X_2 has been taken. Everyone to the right of X_{12} prefers X_2 to X_1; everyone to the left of X_{12} prefers X_1 to X_2. In Figure 1 (a), $F(x)$ represents the distribution of first choice position by the electorate with the policy alternatives arrayed such that any voter's entire preference ordering can be inferred in terms of distance from this point to any other. By assumption, $F(x)$ is a normal distribution. In this model we are allowing

*a somewhat stronger assumption than single-peaked preferences

for abstentions from voting. Anyone having a preference intensity less than a designated threshold level is assumed to abstain. Let this threshold rather equal the distance PI_0 in Figure 1 (b) corresponding to the alternative distances $\pm \triangle X_0$ around X_{12} in Figure 1(a). This means that alternatives separated from one another by no more than PI_0 cannot be distinguished preferentially by any voter.

A knowledge of the aggregate vote outcome — percentage of votes for X_1, X_2, and abstentions — together with the positions of X_1 and X_2, the order of the array of alternatives that enables individual preferences to be inferred from first place positions, the assumed variance of the first choice distribution and the size of the threshold preference intensity PI_0 (with its corresponding $\pm \triangle X_0$), make it possible to infer a particular $F(x)$ and a cutoff point X_{12}. If abstentions are zero ($PI_0 = 0$) everyone to the right of X_{12} votes for X_2, everyone to the left of X_{12} votes for X_1. With $PL_0 > 0$, as in Figures 1 (a) and (b), voters with first choice positions in the interval $X_{12} \pm \triangle X_0$ abstain.

The basic relationship is that preference intensity relative to X_1 and X_2 for any voter is: $PI^i = |\ X_1 - X_i\ | - |\ X_2 - X_i\ |$. Any voter with preference location to the left of X_1 or the right of X_2 experiences a preference difference between X_1 and X_2 equal to $X_2 - X_1$. Only voters located in the interval between X_1 and X_2 have preference intensities (for either X_1 or X_2) other than $X_2 - X_1$ — and for these it is less than $X_2 - X_1$. At X_{12}, $PI1^i = 0 = PI2^i$, and PI^i rises toward $X^2 - X^1$ as the preference location approaches either X_1 or X_2. For locations within the interval $X_{12} + \triangle X_0$, $PI^i = PI_0$.

Thus, Figure 1 (b) translates PI relative to X_1 and X_2 as a function of preference location and Figure 1 (c) uses both these materials to express the distribution of preference intensities.

The diagram and their underlying relationships indicate the following:

(i) A normal distribution of voter preference locations does not generate a normal distribution of preference intensities, as the various approaches seem to require.

(ii) The 'cutoff' alternative X_{12} inferred from actual voting does not establish a unique pair of locations for X_1 and X_2 on the X axis but is in fact compatible with an infinite number of pairs of locations for X_1 and X_2. Each pair, however, leads to a different preference intensity distribution because it leads to a different relationship between PI^i and individual preference locations. So the dashed lines in Figure 1 (a) — with X_1' and X_2' as a different pair of alternative locations — establish the dashed relationship between PI^i and X in

Figure 1 (b), and thus lead to a different frequency distribution of preference intensities in Figure 1 (c) (G' instead of G). Actual voting distributions do not establish unique mean preference intensities for the group, even where the type of distribution of individual preference location is assumed.

(iii) In general, preference intensity is not measurable as simple distances in alternative space; distances in this space translate into preference intensities only when the location of the voted-on alternatives in this space is specified (or determined), and this is a consideration typically excluded from the kind of approach being considered.

(b) Second, the asserted procedure of this type of approach is to partition the voting population into different groups, for each of which it is assumed that the group is distributed normally with the same variance. This is questionable:

(i) Such a partition may exist for one specific set of alternatives (e.g. X_1, X_3, X_5 ...), but it will generally differ for other sets (e.g. X_2, X_4, ...).

(ii) Most natural ways of classifying subgroups will result in some groups that are not normally distributed, but heavily clustered, because the choice context is more closely relevant to them than to others. Typical classifications create groups with differing degrees of partiality vis-à-vis a given set of alternatives. Intra-group distributions are likely to vary greatly.

(iii) To generate groups for which a normal distribution holds is in fact an artificial — and difficult operation. To generate a whole partition of such normally distributed groups would generally require randomization. Even randomization is not a sufficient condition for generating a partition of normally distributed groups under some choosing situations.

(c) This type of approach clearly is beset by many pitfalls. It seems a better strategy to measure preference intensitites for different individuals — *and use these as individual weights* — rather than to derive differences in mean group preference intensities, to weight the preferences of *groups*.

5. In an earlier version of Tideman's paper, he argued that Edward Clark's proposed procedure for eliciting honest marginal valuations from the electorate was defective. In fact, it is Tideman's criticism that is mistaken. This issue illuminates an important aspect of the real structure of Clark's procedure, one which is not stressed in the revised version, and for this reason is worth considering here despite correction

of the error itself.

Tideman's original criticism was that under Clark's procedure each voter is virtually made 'a monopolist seller of changes in the levels of public activities', and therefore can be expected to 'sell each unit at a price that gives him the whole surplus from it. One cannot expect honest revealed preferences from this system.'

It is not true that each voter has a monopoly entitlement to the consequence of his abstention. The structure of the procedure is more complex than that: each participant is bound in a web of reciprocal relations with others, and does not *independently* determine anything. The possibility of marginal abstention by each is bound up with the other participants' reciprocal, conditional abstention. All participants in effect face reciprocal expectations about behavior of one another — only when they are all consistent does anyone get what he wants. This is one aspect of the situation which makes it like a private competitive equilibrium. It is the absence of any such interpersonal contingency that absolves each voter from responsibility for the consequences of his choice in the pure public good case — that abstracts the individual's behavior toward public benefits from the true personal opportunity cost of that behavior. Clark cleverly constructs the link by creating an artificial web of interpersonal contingency. Whatever its defects, its real achievement is in making each individual *more* closely interdependent with his fellows in the process of public choice, not — as insular monopolist — less.

6.   The foregoing notes suggest a broader comment on preference misrepresentation. Both Babb and Bohm report experiments in which the observer expected considerable free rider misrepresentation by the participants, but failed to find it. Specific difficulties can, of course, be pointed out with these experiments. But despite them the salient negative finding suggests that researchers may have to make a deeper effort to try to understand the situational aspects of choice. Combinations of uninformedness, risk, large numbers of participants, the presence of difficult calculations (especially where a group choice procedure calls for a sequence of actions rather than a single stage choice), and other such particular situational features, may make strategic misrepresentation considerably less important in practice than abstract analysis would suggest in principle. The discrepancy between the two is certainly relevant to how public choice ought to be made.

This negative warning echoes Terry Clark's relaxed and optimistic view of our problem in this area. These days one is well-advised to grasp at almost anything that is relaxed and optimistic.

The Democratic Response of Urban Governments: An Empirical Test with Simple Spatial Models

Wayne Hoffman

INTRODUCTION

Among the normative concerns of social scientists, one of the greatest lacunae is evaluation of government performance in light of populist democratic criteria. While formal theorists have continued elaborating models of a competitive political process, and while the search for an appropriate and serviceable welfare criterion goes on, the empirical application of our existing descriptive and prescriptive insights mostly waits in abeyance. This paper offers one attempt to apply formal criteria to the evaluation task. We first briefly develop several low-level, admittedly non-complex, statements about the *expected* outcome of public sector activity according to the median preference model. The model's descriptive statements then serve as evaluative and comparative criteria as we test its adequacy under several assumptions concerning democratic responsiveness. Finally we attempt to explain variations in the degree of responsiveness exhibited by ten American urban governments by reference to three aspects of political structures.

The value of our approach emanates, we believe, not from any new insight into the formal interpretations and predictions provided by positive political theorists.[2] Rather, we consider the approach a nascent attempt to match formally derived expectations to non-experimental data from both public opinion surveys and aggregate characteristics of large American cities. Our route allows us to link micro-level responses of citizens to macro-level behavior of governments to assess the

---

For background to work on this project see Note 1.

responsiveness of urban governments. We find in the existing literature no similar attempt at empirical application of spatial models or evaluations of urban government performance.

A SIMPLE EXPOSITION OF DOWNS' MODEL

The median preference model elaborated by Anthony Downs offers one mechanism for relating the distribution of citizen preferences to an expected level of government policy output.[3] The general notions of *An Economic Theory of Democracy* are so widely known that we here summarize only a few crucial points. Ostensibly Downs seeks to explain the behavior of political parties — teams of self-interested men with a common purpose — and of citizens in a democratic polity. His illustrations are clearest for political systems with only two competitive parties. The principal postulate of the theory is that party members have as their chief motivation the desire to obtain the intrinsic rewards of office: 'The main goal of every party is the winning of elections.'[4] The actions of a party are thus aimed at maximising votes.[5]

To maximize votes, parties formulate platforms which take positions on public policies that closely approximate the median of the preferences held by the citizenry. Frequently this is referred to as the 'position of the median voter'. If either party took a policy position other than the median, the other could move closer to the median, thereby assuring itself more than fifty percent of the votes and win the election.

This simple model of party behavior suggests that the *policy actually-pursued* by the party once in office should be very close to that preferred by the voter with the median preference in the election.[6] Downs stops short of stating this explicitly. However, he lists this prediction as the first testable proposition which may be deduced from his theory.[7] Over time, the government's policy should reflect median voter's preferences as well even if the teams of men holding office change from one election to the next.[8]

THE EXPECTED OUTCOME OF GOVERNMENT ACTIVITY

A simple extension of the model from this point allows us to specify an expected outcome of government activity in terms of the distribution of citizen policy preferences.[9] We need first, however, a simple way of characterizing 'policy': Downs assumes that alternative policy *options* may be arrayed along a single dimension. We shall assume that such a dimension may be represented by the amount of public expenditures devoted to the policy. With typical examples of local government activity – for example, the provision of police protection, or local streets and trash collection – it is not uncommon to discuss improving or cutting back on the services by reference to an increase or decrease in budget outlays.

Such simplication undoubtedly does injustice to the complex character of public policies. Other important dimensions include the *manner* in which public employees perform their duties, the *aesthetic appeal* of public facilities, or the *location* of new roads, schools and libraries. But we must begin somewhere. Our simplifying assumption is not inconsistent with that of Downs who characterized government policy in terms of *budget decisions:* 'According to our hypothesis, governments continue spending until the marginal vote from additional expenditures equals the marginal vote loss from financing.'[10]

In terms of the single expenditure dimension for a given policy, we can specify our most general hypothesis:

> *Hypothesis:* The policy pursued by a responsible government representing the median voter's preference should generate in the (appropriate) population a distribution of preference *equally divided between those who prefer more spending on the policy and those who prefer less.*

Under ideal conditions, we should be able to obtain such information on the entire array of policies pursued by a government. In practice in the empirical analysis that follows, we will focus on several of those policies most commonly accepted as within the 'scope of public sector' and which taken together account for a significant portion of the public expenditure commitment of American local jurisdictions.[11]

THE OPERATIONAL MODEL

Departures of the observable from the expected outcome offer an explicit measure of the degree to which local government policies depart from democratic criteria implicit in the Downs model. We operationalize such departures by calculating the difference between the percentage of the surveyed population who desire increased spending and the percentage who desire decreased spending on a government policy. The measures, which we refer to as 'percentage difference' (PD) scores, may extend from +100 to -100 although the actual data is much more limited in range. PD scores greater than zero are indicators of 'excess positive demand' suggesting, according to the model, that governments are 'under-producing' the policy. Negative scores indicate 'excess negative demand' meaning the policy output level is too high. Government output in accord with the median preference standard should result in a PD score close to zero.

The PD scores provide two dimensions for comparison. One comparison is among the types of policies. We may find that there is a consistent pattern of over-production or under-production of some types of local government policies. The second comparison is among the ten units of government. The difference in PD scores across these 'polities' is of particular interest for it provides an opportunity to examine several conditions under which governments respond in accord with the model. Three such conditions will be spelled out below.

Before discussing our survey data, two other sources of *preference imbalance* (PD scores different from zero) should be mentioned. First is the use of survey data itself.

Survey data has been criticized because it fails to account adequately for the level of respondents' information, intensity of preference, and for lack of constraint on expressed demand when public expenditures are involved.[12] Our analysis illustrates some of these difficulties. Fortunately, we do not have to treat our measures as absolute numbers. We can compare the magnitude of the survey measures across policies and units. In addition, we attempt to identify differences in results due to differences in question format. Specifically, we compare questions that elicited from all respondents their preferred spending position on the ten policy items with a more open-ended format which allowed the respondent to select policy items salient to him. The latter, in most instances, produces smaller PD imbalances which are more in accord with expectations of the model.

Secondly, in formulating our major hypothesis, we parenthetically referred to the distribution of preference outcomes in an 'appropriate' population of the citizenry. The simplest expectation is that all potentially eligible voters form the appropriate population. Downs discusses several incentives that parties might have to weight the preferences of some citizens more than others. Unequal weighting of votes is a departure from the populist democratic criterion implicit in selecting the median preference as the collective choice. Our operational model permits us to observe the effect on PD imbalance of different weighting assumptions. Specifically, we explore the possibility that political entrepreneurs differentially weight the preferences of more active political participators, racial minorities, and wealthier individuals. In each weighted version of the model, some PD imbalance is observed. Where weighting produces less imbalance however, it suggests departures by the governments from the simple democratic model.

DESCRIPTION AND INTERPRETATION OF THE SURVEY DATA

The data on the distribution of preferences is based on representative samples of residents surveyed in *each* of ten large American central cities. A major part of the 1970 survey questionnaire included information on both local leaders' and citizens' policy-related concerns. General evaluation of governmental performance were also elicited.[13]

Of particular use for our purposes were items that posed the alternative of increased spending, decreased spending or spending at the existing levels for a set of ten municipal services. The particular form of the question, under certain assumptions, can be used to place the respondent in the preference distribution of the city's population. The question read:

> Here is a list of services and problems. Some we have talked about already, others we have not. For each I want you to tell me whether you think the local agencies should spend *more money, less money,* or *about as much money* as is now spent on those services, or problems. Remember, that to spend more on something, the local government either has to spend less on something else or it has to raise taxes.

A list of local policies followed and the respondent was requested to indicate his preference for each item.[14]

The position is a relative one depending both on the city's existing level of service and on the respondents' perception of this level. We assume the respondent has a rough idea of the tax price he pays and of the service provided and that he seeks to maximize his welfare surplus, or reduce his welfare loss, by stating his preferred movement in government spending. Unfortunately, we cannot identify the exact median position in the preference distribution. We approximate it by considering these points:

(a) 'the center' — satisfied with the existing level of service;

(b) a point in one tail of the distribution for which an *increase* in the service level would reduce the individual's welfare loss by moving him closer to this most preferred position; and

(c) a point in the other tail of the distribution whereby a *decrease* in the service level would reduce his welfare loss.

The number of persons who fall at each of these three points generates a preference distribution similar to Figure 1.

From the set of policy items, we selected ten of particular salience to the respondent and which represented substantial expenditure

Figure 1. Example of Policy Spending Preference — Distribution
from the Survey Data*

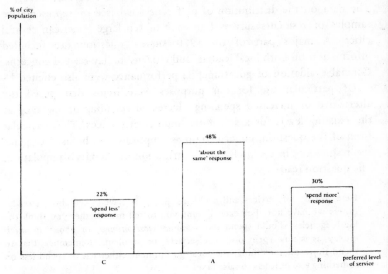

*PD score imbalance in this example would be (30 - 22) = 8

commitments in most large cities. Below we report results on all ten
items separately as well as summary results for four distinct policy
groups. The complete rationale for this clustering of items has been
detailed elsewhere.[15]

The first two items concern 'air pollution control' and 'improving
mass transit'. These we have termed the 'environmental services' group.
The next group of four — our 'traditional services' — includes police
protection, street building and repair, installation of street lighting,
and trash and garbage collection. Seventh is spending on public schools,
or 'educational services'. Three 'welfare service' items — building low
cost housing, providing medical care for those unable to afford it,
and public assistance (like AFDC) — form the fourth policy group.

The first two environmental service policies, we reasoned, are
relatively new to the agenda of most cities. Neither citizens nor
political parties may be fully attuned to the demand for a clean
environment. (This was more likely the case in 1970 than today.)
Part of the imbalance for these policies may derive from their new-
ness. While police protection received strong national attention with
widespread campaign rhetoric favouring law and order, we expect that

Table 1. Summary of Preference Distributions in Ten Cities for Ten Policy Items (PD Scores for Total Adult Population)

	Pollution control	Mass transit	Police protection	Street lighting	Street repair	Trash collection	Public schools	Low cost Housing	Medical care	Welfare	Mean of city scores
Boston	59	41	69	30	51	19	66	61	59	21	48
Baltimore	33	38	65	33	41	24	67	58	72	46	47
Atlanta	53	47	58	36	56	49	62	51	62	45	52
Nashville	•47	19	62	34	60	31	60	43	61	34	45
Milwaukee	47	27	42	21	31	14	30	34	50	8	29
Kansas City, Mo	40	34	64	36	62	37	57	30	55	27	44
Kansas City, K	23	15	53	33	49	23	53	45	63	31	35
Denver	58	59	47	28	42	14	44	40	41	-1	37
Albuquerque	40	19	46	33	48	36	52	34	41	21	37
San Diego	59	30	31	32	24	10	38	26	36	6	29
Mean of Absolute Scores	46	33	54	45	31	25	53	42	54	26	41

the four 'traditional service' policies should still be close to the preference of the median local voter. PD scores should be near zero.[16] Public education is complex in that some multiple purpose governments are not responsible for its financing (true for four of our ten cities). Often, however, the single purpose government — for example, school district — is nearly coterminous with the multi-purpose jurisdiction. This mitigates the problem of preference revelation. (In fact tax bills in many cases come in the same envelope.) Thus, we expect the PD scores for public schools to be close to zero as well.

For the welfare policies, our expectations about PD imbalance are mixed. Highly redistributive and expensive, these policies should be salient to rich and poor citizens and to political parties. Local governments have numerous options for construction of low income or 'public' housing; the policy should thus reflect local preferences even though funding may be from higher levels of government. For the other two welfare services — public assistance and medical care — the policy is highly dependent on federal, state and county decisions. Preferences of the local citizenry may thus not be reflected in local welfare policies. The items provide an interesting comparative test for their PD scores should be more imbalanced than those for the other policies.

OUTCOMES — THE DISTRIBUTION OF CITIZEN POLICY PREFERENCES

Table 1 presents the data from the first application of our model to the observed distribution of policy preferences. The results are quite surprising. The numbers shown are the PD scores for *each* of the ten policy items in *each* of the ten cities. According to the major hypothesis of our model, these PD measures should be close to zero. That is, if a policy were produced at the level preferred by the median voter, we would expect many citizens to respond that the government should 'spend about the same'.[17] And the proportion saying 'spend more' should approximately equal the proportion saying 'spend less'. The high PD scores show that this expectation is poorly met. The scores are (with one minor exception) all positive — representing very large excess positive demand. They range from a high of 72 for medical care in Baltimore to a low of -1 for public assistance in Denver.

Consider the distribution of preferences that underlies the Boston scores in the table. For each policy, a far greater number of the voting age population[18] prefer an increase of policy spending than prefer a decrease. For air pollution control, this difference in proportions is represented by a PD score of 59.

There are several interpretations for this finding but the most general implication is that the government's expenditure on air pollution control as perceived by the citizenry should be substantially increased. Exactly *how much* additional expenditure should be allocated cannot be determined from our data. The median preference model would predict that parties seeking to win the next election should take a strong position on increased expenditure on pollution control.

The first row of the table shows that the policy most under-produced in Boston is police protection. Sixty-nine percent more of the citizenry prefer increased spending for police than prefer decreased spending. The least under-produced policy is trash and garbage collection, with a PD score of only 19. The far right figure in the first row (48) is the mean (absolute) PD score for Boston for all ten policy items.

While the table provides data for ten cities and ten policies, we cannot discuss each in depth. Our purpose is examination of the model's applicability under a variety of conditions. As a summary of the table, the column on the right presents average PD imbalances for each city. Milwaukee and San Diego score 'best' with a low of 29. Atlanta receives the 'worst' and highest score of 52. The bottom row of the table provides an indication of the average relative imbalance among policies. These means of the absolute values of the scores show that public assistance is the policy *least* under-produced in the ten cities.[19] As in Boston, on average the policy *most* under-produced is police protection.

Rather than commenting on differences among the ten policies, we refer to Table 2 which combines policies into the four groups previously defined. The first four columns show the average PD scores in each city for environmental services, traditional services, education and welfare service policies. One rationale for such a policy grouping is that it provides a less demanding expectation of local government responsiveness.[20] Parties may not respond to each policy item in the budget, but to broader arrays of citizen-policy concerns. Combining several closely related policies operationalizes this broader conception of responsiveness. The averages by policy group at the bottom of Table 2 do show much less variation than do the ten policies of Table 1. The specific results are nevertheless unanticipated. Average excess demand for education (PD = 53) is the highest of the four, not only on average but in each city. Our expectation was that environmental policies — as new issues — and welfare services — because of extra-local influence — would reveal the greatest imbalances. We consider

Table 2.   Summary of Preference Distributions in Ten Cities for
Four Policy Types and for Tax Levels (PD Scores for
Total Adult Population)

	Environmental services	Traditional services	Education services	Welfare services	Taxes
Boston	49	42	66	47	62
Baltimore	35	41	67	59	65
Atlanta	50	50	62	52	27
Nashville	33	47	60	46	9
Milwaukee	37	24	30	30	62
Kansas City, Mo	37	49	57	37	22
Kansas City, K	19	39	53	46	44
Denver	58	33	43	28	32
Albuqurque	29	41	52	32	14
San Diego	44	24	38	22	21
Mean of Absolute Scores	39	39	53	40	36

alternative measures before commenting further.

Part of the explanation for 'under-production' imbalances is
suggested by the tax imbalance score in the right hand column of
Table 2. The tax question in the survey asked the respondent: 'From
what you know, do you think taxes are too high, here in (CITY)
about right, or are they too low to pay for needed services?' The
results were striking: non-missing responses for all ten cities combined
were 'too high' 46 percent, 'about right' 44 percent, 'too low' about 10
percent. Substantial numbers of citizens perceive existing
taxes as too high. The pattern is differentiated somewhat by city,
however, as shown by the tax PD scores, calculated by subtracting
the percentage who said 'too high' from the percentage responding
'too low'. The average tax PD score is 36, with the lowest scores for
Nashville (9) and Atlanata (14). The consequences of tax constraint
for imbalances are considered in revised models below.

One reading of the data in Tables 1 and 2 might be that local
governments in these cities are quite unresponsive to median demand.
Should we speculate that local political entrepreneurs base their
decisions on preferences of some unrepresentative subset of citizens? [21]
Despite under-production for almost all items for all cities, we might
seek important differences among cities to explain the rather marked

variations in non-responsiveness. We return to city differences shortly. First, we explore some vagaries of the survey data that may contribute to the imbalanced distributions.

In some modifications of the median preference model, as well as in more traditional normative concerns, the expectation is that democratic governments often respond to preferences intensely held by minorities rather than to weakly held preferences of majorities.[22] In the voting process it is unclear precisely how intensity of preference enters. Presumably since candidates and parties offer stands on multiple issues, a citizen's vote reflects his choice of party based on perceived differences in platform on one or a few issues of greatest concern rather than on the entire range of issues. In the ballot booth, at any rate, the vote of the semi-apathetic is equal to the vote of the impassioned.

The survey questionnaire, unfortunately, did not pay careful attention to such intensity of response. However, one follow-up question to the preference items did ask the respondent to rank up to three of his most preferred policies for increased local government spending and up to three of his most preferred candidates for lesser government effort. The PD measures in Tables 3 and 4 incorporate these responses. They are computed by including up to six policies in which the respondent reported greatest (favorable or unfavorable) interest.

The imbalance remains marked, but decidedly less so than in Tables 1 and 2. In Table 3, five policies, including three traditional services, show very low imbalances. And negative PD scores for about half the cities indicate *over-production*. On the other hand, schools, police and medical care are substantially under-produced in all cities.

The right margin of Table 3 again summarizes the scores for each city.[23] Cities differ less than in Tables 1 and 2, although Milwaukee and San Diego still show low imbalances.

At the bottom of the Table are two *summary* measures that should be conceptually distinguished. The mean *net* score (average of PD scores including positive and negative values) reveals which policies tend to be under- or over-produced and by how much. Public schools replace police as the single policy most typically under-produced, but police and medical care rank a close second and third. By contrast, the bottom row of the table shows the mean of the absolute values of the PD scores. While closely related to average net scores, the measure better reveals the degree to which individual policies are produced in accord with the median model — irrespective of over- or under-

Table 3. Summary of Preference Distributions in Ten Cities for Ten Policy Items
Most Salient Policies Only (PD Scores for Total Adult Population)

	Pollution control	Mass transit	Police protection	Street lighting	Street repair	Trash collection	Public schools	Low cost housing	Medical care	Welfare	Mean of city (Absolute) scores
Boston	0	- 9	41	-10	- 0	-10	33	12	27	- 6	15
Baltimore	-19	-13	30	-15	- 2	- 5	40	13	33	7	18
Atlanta	- 6	-12	20	-13	- 1	9	39	1	22	6	13
Nashville	- 6	-28	29	-16	10	- 3	47	1	31	3	17
Milwaukee	14	-11	27	- 6	- 5	- 5	15	7	29	- 4	12
Kansas City, Mo	-10	-15	40	- 1	11	4	37	-14	31	- 3	17
Kansas City, K	-21	-26	32	- 5	7	0	36	- 3	24	4	16
Denver	28	20	18	-12	- 2	- 3	25	3	19	-19	15
Albuqurque	7	-26	19	- 4	7	4	40	- 7	16	- 0	13
San Diego	29	-10	12	- 0	- 5	- 5	25	-13	18	-13	13
Net Mean	2	-13	27	- 8	2	- 1	34	0	25	- 3	
Mean of Absolute Scores	14	17	27	8	5	5	34	7	25	6	15

production. This measure should be compared to that at the bottom of Table 1 to obtain an idea of the improvement offered by this model where measures of most and least salient policies are used.

Table 4 presents the PD scores by four policy areas, again including only the 'salient' questions. Education still stands out, even more sharply, as the policy most under-produced in each city (except Denver) and on average. The summary mean absolute scores at the bottom indicate that traditional services are the policies produced

Table 4. Summary of Preference Distributions in Ten Cities for Four Policy Types **Most Salient Policies Only** (PD Scores for Total Adult Population)

	Environmental services	Traditional services	Education services	Welfare services
Boston	- 8	16	33	22
Baltimore	-27	6	40	36
Atlanta	-15	13	39	19
Nashville	-26	14	47	23
Milwaukee	1	9	15	22
Kansas City, Mo	-22	32	37	11
Kansas City, K	-35	21	36	19
Denver	31	2	25	1
Albuqurque	-15	17	40	7
San Diego	13	0	25	- 5
Net Mean	-10	13	34	16
Mean of Absolute Scores	19	13	34	17

most in accord with median preferences when intensity of preference is taken into account. The mean net score suggests that environmental services are generally over-produced. This result runs directly counter to our expectation and to the finding in the first two tables.

Our analysis to this point provides several interesting findings in light of the median preference hypothesis. Methodologically, empirical support for the general model is greater using responses that incorporate salience of the issue or intensity of preference. Substantively, we find that most policies are *under-produced* using the model as the standard of comparison. Few of our expectations about specific *types* of policies are borne out. This seems in good part due to the absence of an adequate budget constraint in the questionnaire format — since the tax preference item showed that respondents felt taxes were too high. Clear differences remain among cities and policies.

ALTERNATIVE MODELS — WEIGHTING CITIZENS UNEQUALLY

Thus far the analysis has assumed that political parties weight equally the preferences of all potential voters. Application of the model under this assumption generates substantial imbalances for all policies. Alternative assumptions about the calculus of party position-taking are possible; five will be examined here. The first alternative model weights the preferences of individuals in the sample by our best estimates of the likelihood of political activism.[24] While Downs argues that one reason for 'rational' abstention from voting is that the citizen perceives no difference in party positions, other factors also contribute to non-participation. Some urban residents are likely not to vote in local elections even where party positions are clear and personally relevant. We reasoned that their preferences would be of least concern to political parties. Only the most frequent voters — roughly 50 percent of the sample — were included in this first alternative model.

The second alternative model also follows Downs by refining his theory about party position-taking. Citizens are weighted in direct proportion to the private sector resources they possess.[25] In operational terms, for example, responses of individuals in our survey with a $12,000 annual income are weighted twice those with incomes of $6,000. If this model should produce lower PD imbalances, we would have some evidence that private sector resources are translated into public sector outcomes. The result would run counter to the normative concerns of many political theorists, ordinary citizens, and advocates of political equality who prefer that government policy benefits be effectively distributed in a more equal fashion than income. However, the result would not be altogether surprising. A great deal of empirical research has found that political participation — generally assumed to affect policy outcomes — is quite predictably and positively related to income and social status.[26]

The third alternative model builds on the economic concept of 'effective demand'.[27] Demand for, and consumption of, private commodities is limited by the individual's budget constraint. As income rises, so does demand for goods (in the normal goods case). In the public sector, votes on policy expenditure are not so directly constrained by income. The voter may assume (1) that other policy expenditures in the budget will be reduced and reallocated; *or* (2) that under existing tax laws, others may disproportionately pay for his preferred policy increases. The high excess positive demand and the PD

Table 5. Summary of Median Preference Response for Eight Weighted Models

Policy Area	Adult population (1)	Participation (2)	Income (3)	Effective demand (4)	Minority −1 (5)	Minority −2 (6)	Combined −1 (7)	Combined −2 (8)
Taxes	36	34	31	31	34	32	23	25
Absolute Means for Salience Measures								
Environmental	19	19	17	18	18	16	16	16
Traditional	13	14	14	11	13	12	12	12
Education	34	34	37	34	31	31	34	36
Welfare	17	14	12	13	15	12	12	12
Absolute Means for Original Measures								
Environmental	39	40	40	36	40	41	39	38
Traditional	39	39	37	34	38	36	31	33
Education	53	53	54	50	51	49	48	50
Welfare	39	37	32	33	38	34	23	26
Mean for all Ten Policies	41	40	38	36	39	37	32	33
Net Means for Salience Measures								
Environmental	−10	−7	−5	7	−8	−3	5	2
Traditional	13	14	11	11	13	12	11	11
Education	34	34	37	34	31	31	34	36
Welfare	16	12	6	10	13	10	−3	−1

tax imbalance in Table 2 are strong evidence of this tension between preferences for increased expenditures and unwillingness to pay tax increases.

In the aggregate (barring free use of extra-local governments' resources), expenditures are constrained by revenues. For local governments, increased expenditures imply increased taxes. One calculus that parties might follow to simplify this dilemma would be to weight simultaneously the expenditure preferences of individuals by the probability of their willingness to pay increased taxes. Preferences of citizens favoring increased expenditures, but opposing tax increases, would be discounted.

To operationalize this criterion we calculated 'effective demand' scores for each respondent. Respondents were excluded from the model if their expressed preferences for increases on all policy items were above the mean for the entire sample *and* their tax preference indicated they strongly opposed tax increases.[28]

The two final models concern the presence of Blacks in urban political arenas. They build on the hypothesis that preferences of Blacks would be discounted by white politicians. In version 1 of this model, we excluded the black population of each city for the calculation of PD scores. In version 2, white and black respondents were weighted by their respective proportions of the city's population. For Baltimore, as an example, Blacks comprised 46 percent of the city's residents in 1970, and their responses were weighted 0.46; correspondingly, responses of Whites in Baltimore were weighted 0.54.

Table 5 summarizes the results from all five alternative models. It shows the mean of selected *absolute* and *net* PD scores for all cities by policy areas and for tax preference. The left column shows the mean PD score for the first weighting scheme for the total adult population — results from Tables 1-4. The column serves as a standard of comparison for the PD scores generated by the models in columns 2-7.

The alternative schemes have remarkably little effect on the PD imbalances; they are reduced for only a few policy measures in each model. The participation model mainly reduces the imbalances for welfare, but only from 2 to 4 precent. The income model also alters welfare imbalance, environmental policies and the tax measure, but only slightly. For some policies (for example, education), the imbalance is actually increased (although this may be due to rounding errors).

The possibility that each weighting scheme was appropriate to only one or two policy areas prompted us to examine two final 'combined weight' models. The results are in the far right columns of Table 5.

Combination 1 multiplicatively used the weights of models two, three, four and five. Combined model 2 is the same except the proportional weighting of Blacks in model six is used instead of the weights of model five. The most notable effect of the combined models (compared to the unweighted column one) are the reductions in the tax preference PD scores, and in the traditional and welfare service scores.

A final comment on the figures in the bottom third of Table 5. The PD scores here are lower than those above it since they measure average *net* scores. Even if welfare, for instance, was substantially over-produced in five of our cities (negative PD scores), and under-produced to the same degree in five cities, the average for the ten cities would be zero. As with previous summary measures, the absolute scores are strict criteria for evaluating the model; all PD socres, whether positive or negative, are evidence of deviations from the median expectation. The mean figures for *net* scores on the other hand, provide information on the average tendency toward over- or under-production of the policy.

To summarize the analysis of the weighted models, the results fall far short of our initial expectations. The distribution of preferences in the weighted populations should have produced PD scores closer to zero than the unweighted median preference model. In fact, they move the PD scores only slightly in that direction.

Two interpretations of these findings are possible. One is that the survey data — despite the alternatives considered — are poor indicators of the 'actual' policy preferences. As with all empirical research, measurement error and inadequate operationalization of variables or concepts is most likely part of the problem. A second possibility is more intriguing: the Downs model simply does not aptly describe the decisions of governmental leaders in providing local policies. The imbalanced outcomes might then be taken as evidence for an 'undemocratic' response according to the median preference standard. Recall that Tables 1-4 showed that cities vary substantially in the degree to which they converge to the median; PD scores range from near 0 to the high 60s. Rather than dismiss these results out of hand, we will consider the PD scores as indicators of differences in demo-cratic responsiveness among the cities.

EXPLAINING INTER-CITY DIFFERENCE IN DEMOCRATIC RESPONSE

The analysis which follows is particularly exploratory. It seeks to identify the conditions under which deviations from the median preference are likely. With only ten cases in our sample, the multiplicity of factors which could account for deviations cannot be adequately examined. Just three of the most obvious are described below.

First, we consider the effect of the form of local government. It has been hypothesized that 'reformed' cities, those which modified their political structure to reduce the corruption of machine politicians, are less responsive, especially to poor and minority urban residents.[29] If this is the case, an index of reform should be positively related to the absolute PD imbalances in our cities. Our index of reform is scored 0 to 3 depending upon the presence of common reform structures: at large councilmanic elections, absence of partisan affiliation on the ballot and the assignment of chief executive functions to a city manager.

A second political characteristic is implicit in Downs' theory — the extent of party competition. Where competition is strong, the median preference is more likely to be reflected in local government policy. Where competition is traditionally weak or absent, the party in power should be freer to pursue policies less in accord with the median voter since it has less fear of being thrown out of office in the next election. Adequate measures of local party competition are hard to obtain. We used a combined measure based on the two-party division of the presidential and gubinatorial vote in the city in several years prior to 1970.[30] We predict negative relationships between the degree of competition and PD imbalance in the cities.

A third condition with political relevance that may account for the dramatic under-production of policies is city fiscal capacity. It is hard to operationalize adequately, but the *mean income* of families within each city provides a rough approximation of the tax base.[31] A lower tax base necessarily requires a higher tax rate to produce policies at the level provided in wealthier jurisdictions. Fiscal capacity then should be negatively related, we hypothesize, to responsiveness as represented by our PD scores.

Table 6 shows the zero order correlations of the three conditions of democratic responsiveness with the mean absolute measures for each policy area. These imbalance scores are those derived from our best (Combined 2) weighted model shown in Table 5.

The correlations with the index of reform force a strong rejection of the hypothesis. Cities with reform characteristics are less likely to generate PD imbalances. We interpret this as evidence of greater democratic responsiveness by reformed governments. The other two conditions show several correlations in the expected direction which

Table 6. Correlations of Mean Absolute Measures of Preference Imbalance (PD score for City) with Two Political Structure Variables and One Measure of Fiscal Constraint*

	Index of reform	Party competition	Fiscal constraint
Predicted Relationship	+	-	-
Taxes	-0.32	-0.64	-0.57
Salient Measures (Mean Absolute PD)			
Environmental	0.23	0.49	0.51
Traditional	-0.64	-0.15	-0.31
Education	0.04	-0.27	-0.44
Welfare	-0.65	0.03	0.30
Original Measures (Mean Absolute PD)			
Environmental	-0.18	0.56	0.46
Traditional	-0.23	0.19	-0.11
Education	-0.32	-0.42	-0.52
Welfare	-0.52	-0.60	-0.63
Total Mean PD Imbalance	-0.54	-0.09	-0.28

*Pearson's *r* correlations at 0.70 0.54 0.43 are signficant at 0.01, 0.05, 0.10 respectively.

Total N = 10 cities

Table 7. Correlations of Mean Net Imbalance with Two Political Structure Variables and One Measure of Fiscal Constraint

	Index of reform	Party competition	Fiscal constraint
Policy Area	Salient Measures—(Mean Net PD)		
Environmental	0.66	0.49	0.55
Traditional	0.22	0.08	-0.22
Education	0.04	-0.27	-0.43
Welfare	-0.71	-0.66	-0.54

are statistically significant. However, the results for all policy measures are not sufficiently strong or consistent to provide firm support for the hypotheses. They must be viewed as very tentative evidence of the importance of these conditions for democratic responsiveness.

Table 7 provides additional information on the effects of reform, competition and fiscal capacity on policy outputs. The PD imbalance measures in this table are the net scores (from the bottom portion of Table 5 for the weighted Combined 2 model). These correlations reveal the tendency toward over- or under-production of each policy type. Reformed government has little consistent effect with the exception of welfare services. The negative relationship means that reformed cities are more likely to over-produce welfare services than are unreformed cities. The same is true for party competition. An alternative way of summarizing is to say that governments in the presence of competitive parties are *less likely* to under-produce welfare services than governments in cities where competition is weak. Competitive party systems however are *more likely* to under-produce environmental services relative to the median preference on salient policy items.[32]

The effects of fiscal capacity are quite similar to those of party competition. Cities with higher tax bases are less likely to under-produce welfare services. The reason for this relationship is unclear since most funding for welfare comes from higher levels of government. The significant positive correlation with environmental services is puzzling. In general, environmental services command relatively small proportions of budget outlays, so perhaps the fiscal constraint is not relevant. Only a multivariate model with a greater number of units for analysis could unravel the correlations that our measure of fiscal capacity may have with other variables affecting the provision of environmental services.

In summary, our brief exploration of three conditions that may account for variations in democratic responsiveness among cities produces two expected and one unanticipated result. However, several correlations are weak or statistically insignificant. There is a special need for future research of this sort to consider a greater number of the conditions under which democracy is expected to flourish in the urban context and to pay more attention to adequate measurement of such variables.

CONCLUDING REMARKS

In this paper we have attempted to illustrate a rather obvious — but never before reported — empirical application of simple spatial models describing the behavior of democratic governments. Extending Downs' theory, we predicted that an observable distribution of citizen preferences for policy spending should be characterized by equal proportions of those who prefer more spending and those who prefer less. The results of the analysis reported here are tentative. They suggest that urban governments produce municipal services at levels quite at variance with that preferred by the median voter. Our data show that most policies are 'under-produced' relative to that standard, but this finding is based on analysis that does not adequately consider the political constraints imposed on governments by the citizenry's common resistance to tax increases. Our *weighted* models suggest that for several policy areas, political parties in power do *not* respond equally to the preferences of all citizens. Urban residents who have greater financial resources, who are more likely to be politically active, who express 'effective demand', or who are not black or members of racial minorities seem to have greater influence. However, this variation in influence indicated by our models is very slight.

Treating our models as standards of evaluation for democratic responsiveness, we found that governments with 'reformed' political structures are more likely to be responsive, contrary to the conclusions of earlier research, that party competition enhances responsiveness, and that fiscal capacity constraints limit responsiveness.

The final value of our attempt is its possible contribution to recent normative concerns of social scientists. It seeks to use reliable and unbiased descriptive analytic techniques to evaluate both the policy and the performance of collective choice institutions in our society — in this case the purportedly democratic urban polity.

NOTES

1. Work on this project has been supported in part by grant HD58916-01 of the National Institutes of Health, Department of Health, Education and Welfare, Washington, D.C. The earlier version of this paper presented at Meetings

of the Public Choice Society, March 1975 was entitled 'Subjective Evaluation of Government Performance'. Helpful comments have been provided by Terry Clark, Paul Peterson, and David Koehler. This is Research Report No. 69 of the Comparative Study of Community Decision-Making.

2. For the best comprehensive summary of work to date, see William H. Riker and Peter C. Ordeshook, An Introduction To Positive Political Theory (Englewood Cliffs, N.J.: Prentice-Hall 1973).

3. Anthony Downs, An Economic Theory of Democracy (New York: Harper and Row, 1957).

4. Ibid., 38.

5. The conditions under which parties maximize votes are more limited than Downs' original formulation would suggest. Other rational motives frequently dominate. See Riker and Ordeshook, op. cit., 338-70.

6. The concept of 'median preference' assumes that policy alternatives can be arrayed along a single dimension or continuum.

7. Downs, op. cit. 296.

8. No other 'theory' of democracy so explicitly links citizen preferences to government policies. Frequently, structural criteria — the existence of competing elites or viable political parties, an effective constitution, and formal and actual protection of civil liberties — are used to describe democratic government. Process criteria — widespread participation with equal access to elites, pluralistic bargaining and incremental decision-making — are also common. Both such characterizations may be viewed as conditions under which democratic responsiveness occurs. The Downs model offers an explicit standard for examining the core postulate of populist democracy — the correspondence between mass and elite preferences and policy.

9. Extension of Downs' model to a multi-party system in a multi-dimensional policy space with stragetic behavior by candidates has numerous qualifications. The basic results remain policy equilibrium positions in theoretical work, and demand estimates for government services in empirical work. Examples of the latter are Otto Davis and George H. Haines, 'A Political Approach To A Theory of Public Expenditures: The Case of Municipalities' and Bernard Booms, 'City Governmental Form and Public Expenditure Levels' both in The National Tax Journal, Vol. 19 (1966); and Theodore Bergstrom and Robert P. Goodman, 'Private Demands For Public Goods', American Economic Review' Vol. 63 (1973). Whatever its theoretical validity, the median preference model, we believe, offers an interesting normative standard for assessing government performance.

10. Downs, op. cit., 73.

11. The ten local policies account for over 60 percent of budget expenditures by local governments serving jurisdictions over 50,000 in population.

12. Cf. Terry Nichols Clark, 'Can You Cut a Budget Pie', Policy and Politics' Vol. 3 (December 1974), 3-32.

13. The survey data are used with the kind permission of the Urban Observatory Program of the National League of Cities/The US Conference of Mayors under whose auspices it was collected in 1970 under contract with the Department of Housing and Urban Development, Washington, D.C. Other reports based on this data have appeared elsewhere, including The Nation's Cities (August and November 1971).

14. The exact working of the items: 'public schools, police patrolling the streets street lighting, cleaning and repairing streets, providing medical care to people who cannot afford to pay for it themselves, trash and garbage collection, building low cost housing, controlling air pollution, improve public transportation, welfare and Aid for Dependent Children (AFDC)'. Six other items were deleted as they did not have specific public policy referents, or referred to policy dimensions for which output data were unavailable. Examples: 'ticketing and towing cars' and 'building teen centers'.

15. A theoretical rationale for characterizing local services, with empirical support for distinguishing these four policies, is found in the author's Ph.D. dissertation completed at the Department of Political Science, The University of Chicago, Citizen Policy Preferences and Urban Government Performance.

16. Inter-governmental grants for services like police, sanitation, and street repair might seem to confuse the appropriate level for considering citizen preferences. However, inter-governmental grants are low relative to overall budget expenditures. Further, the median preference model should help explain why governments seek grants on these rather than other services.

17. The statement assumes at least a unimodal distribution of preferences for this policy.

18. The samples were drawn as representative of households within each city. To better approximate the voting population of each city, we weighted the results for each respondent by the number of adults in the household. Because of sampling error and non-response, the distribution of several demographic characteristics did not exactly match the proportion in the 1970 US Census of Population. The discrepancy for three characteristics — income, race and home-ownership — bothered us in particular since we had found in earlier analysis that these were related to policy preferences. We thus weighted each respondent so that the resulting sample distribution for these characteristics matched the census report for the city. The weights changed the preference distribution and PD scores only slightly.

19. In Table 1 we have only used the average of the absolute scores as an indicator of imbalance. Only one average would be altered if we had taken the mean score — that of public assistance.

20. One other rationale for grouping policies is the similarity they share on two dimensions of classification — the 'potential for redistribution' and the 'scope of the policy benefits' to individuals, neighborhoods or blocks, or city-wide. See the author's dissertation.

21. Clark op. cit. reviews an extensive list of problems in using survey data for preference revelation. The large excess demand in the policies reported in Tables 1 and 2 is not unique in survey responses of this sort. One early report of similar response imbalance is Eva Mueller, 'Public Attitudes Toward Fiscal Programs', The Quarterly Journal of Economics, Vol. 77 (1963), 211, for a set of national government policies. Her data allowed her to consider more carefully cost-consciousness of respondents. Specifically, respondents were asked to reconsider their stated preferences in light of increased taxes that might be necessary to finance each program. Mueller then found a substantial decrease in the respondents' imbalance score, sometimes 50 percent or more. While the questions we used were prefaced with the reminder that 'to spend more on something, the local government either has to spend less on something else or

it has to raise taxes', this 'constraint on demand' may not have remained salient to the respondents. This consideration argues for a weighted model as used in subsequent sections. See also note 8.

22. See 'Indifference' and 'Alienation' in Riker and Ordeshook, op. cit., 322-26 for a formal interpretation.

23. The tax score is not shown in Table 4 since the questionnaire included no way to guage intensity of tax views. The only tax imbalance scores is that reported already in Table 2.

24. We had no good direct measure of political activity in the ten city survey file. To estimate the likelihood of political activism, we relied on regression equations from the data reported in Sidney Verba and Norman Nie, Participation In America (New York: Harper and Row 1972). We coded predictor variables in our data in identical fashion to theirs. Equations were estimated using their data for their 'voting participation' measure. The coefficients were then used to create a predicted score for each of the respondents in the ten city survey file. The coefficients were close to those reported in Participation In America, 359. Those below the mean on the voting activism score were excluded from our weighted participation model.

25. See discussion of unequal influence in Downs, op. cit., chapter 10.

26. See Verba and Nie, op. cit., especially chapter 8.

27. See Robert L. Bish, The Public Economy of Metropolitan Areas (Chicago: Markham Publishing 1971) for a simple statement describing and suggesting effects of 'ineffective demand'.

28. We created a combined measure of overall spending preference which ranged from -10 if the respondent preferred decreased spending on all ten policies to +10 if he preferred increases on all ten. The tax preference variable used was based on the respondent's answer to a question about his willingness or unwillingness to have taxes increased if necessary to maintain existing services.

29. See Robert Lineberry and Edmund P. Fowler, 'Reformsim and Public Policy in American Cities' in James Q. Wilson, City Politics and Public Policy (New York: Wiley & Sons 1968), 97-124.

30. Competition is operationalized by subtracting the Democratic percentage of the two-party vote from 100 and taking the absolute value. We performed the calculation for both presidential and gubinatorial votes in each city for three time points in the 1960s and averaged all six scores for the final competition score.

31. Mean income for the city was taken from the US Census of Population for 1970.

32. See Matthew Crenson, The Un-Politics of Air Pollution (Baltimore: Johns Hopkins Press 1971) for a compelling argument that predicts this result. Crenson contends that the collective 'pure public' benefits of environmental policies are least likely to be the 'stock in trade' of active political parties.

Utility and Collectivity: Some Suggestions on the Anatomy of Citizen Preferences

G. David Curry

This paper is an attempt to link certain public choice theories with empirical analysis of citizen preferences. We begin with a reasonably clear, or at least strong, relation: cities with more Irish, Catholic, and foreign-born residents spend more through their municipal budgets. The relation holds across three data sets for American cities: for 51 cities from 1880 to 1968, for 154 cities in 1903, and for 661 cities in 1960.[1] Why? Clark tested and rejected a Downsean interpretation of this relation using data for ten cities and for Boston. His test was based on analysis of questionnaire items concerning preferred spending levels.

This paper presents an alternative conceptualization for some of the same findings, by extending the basic Downsean framework to encompass distinct sectors within a political system. The basic elements of the public choice tradition are well known. We mention only one or two highlights to show how the present paper extends this framework.

Economists concerned with public choice, from Buchanan, Tullock, Downs, and Olsen have largely operated with an assumption of methodological individualism.[2] If Olsen considered how individuals might be led by voluntary organizations to overcome their tendencies to be 'free riders' for public goods, his answer remained consistent with a thoroughgoing individualism: offer more specific inducements to individuals. Most theoretical work in the public choice tradition has continued to use discrete individual citizens as analytical building

This is research report No. 68 of the Comparative Study of Community Decision-Making, supported by USPHS, NICHD, HDO8916-02. It grew out of a paper for a seminar on Collective Decision-Making offered by Terry Nichols Clark and James S. Coleman. Thanks are due to James Q. Wilson and Edward C. Banfield who made the Boston Homeowner's data available to T.N. Clark.

blocks.[3] Frohlich, Oppenheimer, and Young provide a suggestive approach toward an alternative.[4] The dynamic element for their explanation of public good provision is the political entrepreneur. He seeks to make a personal profit by assembling resouces and delivering public services. But for Frohlich, Oppenheimer, and Young, he still assembles resources and delivers services primarily to discrete individuals.

What changes in this analysis if we consider individuals as members of significant social sectors, such as ethnic groups, churches, etc.? Individuals may derive benefits directly as well as indirectly by virtue of sector membership. And there is no reason that utility-maximizing individuals need consider indirect benefits any less important that direct benefits. Indirect benefits have the basic disadvantage of complicating our model. But incorporating them in our model should enhance the model's explanatory power. Such, at least, is our goal.

THE GENERAL MODEL

The utility which an individual actor receives under social policy A may be shown as:

$$U_j(A) = U_j(X_A)P_j(X_A) + f_j(A)r[C(X_A) + C(O_A)] + U_j(X_{SA})P_S(X_{SA})$$

$$+ gj(A)\left\{ h_S(A)r[C(X_A) \cdot C(O_A)] \right\} - D_j(A).$$

Each term deserves comment.

(1) $U_j(A)$ = the utility actor j receives under social policy A.

(2) $U_j(X_A)P_j(X_A) = U_j(X_A)$, the utility which actor j receives from provision of public good X_A, multiplied by $P_j(X_A)$, the probability that the public good will be supplied to j.

(3) $f_j(A)r[C(X_A) + C(O_A)]$ = the fraction (f) of the private goods which individual j receives *directly* via 'contracts' for enactment of social policy A. $[C(X_A) + C(O_A)]$ is the total sum of collection and organizational expenditures which the government must pour back in private goods contracts to supply public good A. $C( )$ denotes simply costs for supplying whatever is shown in parentheses, while O_A is the collective organization, such as a political party, necessary to procure the public good. r is the prevailing profit rate for contracts in the political system.

(4) $U_j(X_{SA})P_S(X_{SA})$ = the utility j receives indirectly from provision of collective good A, by virtue of being a member of sector S.

$P_S(X_{SA})$ is the probability that collective good X_{SA} will be supplied. For example, if j is a grocer in an S ghetto, and A is a policy to increase foodstamps, U_j represents j's utility which derives from the marginal increase in food purchase.

(5) $gj(A)b_S\left\{b_S(A)r\ \ [C(X_A) - C(O_A)]\right\}$ = the utility j receives from that fraction of the profit which other members of j's sector derive from government contracts necessary for providing good A.

(6) $D_j(A)$ = the donations j makes (such as campaign contributions) toward provision of public good A.

The remainder of the paper makes use of this model to structure analysis of citizen survey data. The major sectors (S's) considered are ethnic and religious groups in Boston.

THE BOSTON HOMEOWNERS SURVEY:

The survey data we analyze here were collected under James Q. Wilson and Edward C. Banfield. Respondents are a representative sample of Boston homeowners, stratified by ethnic groups.[5] Our primary focus is on eight items concerning public spending. The public school item illustrates the questionnaire format: 'Do you think we should spend more money on public schools in Boston, even if it costs more in property taxes?' The respondent could answer yes or no, and was then given a probe, 'Why is that?' Most other items followed a similar format: in lieu of 'public schools in Boston' they included 'facilities to handle juvenile delinquents', 'welfare payments to mothers of dependent children', 'the zoo in this city', 'pensions for Boston policemen and firemen', and 'the Boston City Hospital, which is used in great part by low-income people'. Two last items included slight further variations: 'Do you think the state should pay a bonus of $300 to veterans of the Korean War, even if it means an increase in the State income tax?' and 'Do you think that salaries of the mayor and city councilmen should be increased?'

Positive responses to the eight spending items are weakly interrelated. As Table 1 shows, of the 28 intercorrelations, 25 are significant at the 0.05 level. Only one coefficient is negative. Despite the positive intercorrelation among all items, it seems preferable to examine each item separately.

We consider first simple correlations of the spending items with two basic variables, income[6] and religio-ethnic status (see Table 2).

The income-spending relations are weak, with only four of the eight significant. More affluent respondents supported higher spending

Table 1.   Intercorrelation Matrix of Eight Spending Items (Pearson *r*'s).
* = significant at the 0.05 level, two-tailed test

Schools	Schools	JD	Welfare	Vet. Bonus	Zoo	Salaries	Pensions
JD	0.30*						
Welfare	0.12*	0.17*					
Vet. Bonus	0.01	-0.01	0.08*				
Zoo	0.25*	0.14*	0.08*	0.09*			
MC Salaries	0.12*	0.12*	0.09*	0.00	0.19*		
P&F Pensions	0.07*	0.09*	0.18*	0.13*	0.12*	0.15*	
City Hosp.	0.24*	0.21*	0.15*	0.14*	0.18*	0.11*	0.19*

Table 2.   Correlations of Spending Items with Respondent's
Family Income. * = significant at the 0.05 level, two-
tailed test

Spending item	Income
Public Schools	0.042
Aid to welfare mothers	-0.046
Juvenile delinquent facilities	0.106*
Veterans' bonus	-0.174*
Zoo	0.150*
Mayor and council salaries	0.256*
Police/fire pensions	-0.042
City hospital	-0.016

Table 3.   Percentage of Ethnic Groups by Income Category

Family Income	Irish	Polish	Italian	Jewish	Black	Yankees	Total
Under 3,000	8.8	10.7	16.4	8.2	15.3	0	9.4
3,000-4,999	12.4	11.6	13.9	8.2	20.6	0	11.0
5,000-9,999	49.6	56.3	55.0	37.8	45.0	0	44.9
10,000-20,000	18.6	18.5	9.8	31.1	10.7	17.0	19.8
20,000	0.9	1.0	0	7.2	0.8	83.0	8.9
No answer	9.7	1.9	4.9	7.6	7.6	0	6.0

for juvenile delinquency (JD), the zoo, and mayor and councilmen's
(MC) salaries. But they opposed the veterans' bonus.

Income is highly related to ethnicity, as the sample was stratified
to include Yankees only with incomes over $10,000. (No such
income constraint was used in sampling other ethnic groups.) Income
distributions by ethnicity are shown in Table 3.

Is ethnicity more strongly associated with spending preferences than income? Table 4 shows the simple correlations; few are significant. In terms of Clark's findings, we note only that the Irish and Catholics generally do not stand out as favourable towards spending.

Next other variables were considered: neighborhood attachment, organizational membership, political involvement. Simple correlations of measures in these four areas showed that citizens more involved in these activities often favored higher spending . But the relations were weak, and perhaps spurious as these were only simple correlations. A factor analysis was completed of all forty-four variables considered; it showed modertate clustering but by no means sharp factors. For regression analyses, we thus selected individual variables that covered these areas of social involvement as well as certain other variables that showed significant simple correlations with the spending items. About ten variables (different for each spending item) were thus entered as independent variables in a stepwise regression to explain variations in each of the eight spending items.[7] The basic results appear in Table 5.

The weakness of many relations, and the exploratory nature of this paper, do not encourage detailed discussion of results. Still, what stands out?

(1) Perhaps the most general result is that a broad range of standard socio-economic variables explain only a quite limited amount of variance in the spending items. Just three of the eight spending items show adjusted R^2's exceeding 0.05: the veterans' bonus, the zoo, and mayor/council salaries.

(2) Income and ethnicity showed modest relations with spending preferences in the simple correlations. Most of these stand up in the regressions when other variables are added. Specifically, higher income persons favor more spending for juvenile delinquency and for mayor/council salaries. More highly educated persons favor spending for the zoo and oppose the veterans' bonus ($r_{income \times education} = 0.56$). Jews support spending for public schools, juvenile delinquency and welfare. Yankees oppose the veterans' bonus. Catholics support the veterans' bonus. The Irish favored increased pensions for firemen and policemen until a control for working as a fireman or policemen was added (not shown). The Irish then no longer significantly favored pensions.

Table 4. Pearson Correlation Coefficients for Ethnicity and Spending Preferences
*significant at 0.01 level of significance

	Public schools	Juvenile delinquent facilities	Welfare for mothers	Veterans' bonus	Zoo	Mayor/ council salaries	Police/ fire pensions	Hospital
Irish	-0.0023	0.0232	0.0553	0.0952*	-0.0128	0.0281	0.1056*	0.0367
Catholic	-0.1028*	-0.0777*	0.0312	0.1843*	-0.0353	-0.0405	-0.0146	0.0334
Polish	-0.0414	-0.0875*	-0.0447	0.0401	0.0170	-0.0496	-0.0895	-0.0570
Italian	-0.0519	0.0067	0.0665	0.0119	-0.0824*	-0.1131*	-0.0604	0.0055
Jewish	0.1096*	0.1055*	0.0851*	-0.0268	0.0295	-0.0114	0.0480	0.0064
Black	0.0615	-0.0237	-0.0640	-0.0076	-0.0422	-0.0772*	0.0266	0.0075
Yankees	-0.0619	0.0152	-0.0535	-0.2733*	0.0442	0.1911*	-0.0391	-0.0477

Table 5a. Regression of Spending Items on Respondent Characteristics
The table shows the unstandarized regression coefficient *(b)* and the *F* score in parentheses. Using standard assumptions, *F* scores above approximately 2.0 are significant at the 0.05 level of significance; these scores are indicated by an asterisk.

Independent Variables	Dependent Variables — Favorable Responses Toward Spending on:		
	Public schools	Aid to Welfare mothers	Juvenile delinquent facilities
Education	0.002 (0.07)	-0.016 * (2.1)	
Income			0.022 (5.5)
Jewish	0.086 * (4.2)	0.129 * (6.4)	0.155 (10.0)
Catholic	-0.024 (0.5)		
Polish			0.071 (1.0)
Willingness to pay $5 to support neighborhood improvement organization	0.018 * (3.5)		-0.014 (1.4)
Length of residence in neighborhood	0.017 * (2.5)	0.023 * (2.4)	0.015 (1.1)
'In general how do you like living here in this neighborhood? ' (high score = like the neighborhood)		-0.025 * (2.3)	
Knows of 'any neighborhood groups or organizations concerned with improving this neighborhood'	-0.003 (0.1)		
Head of household belongs to (any) club	0.003 (0.1)		
Helped 'in a campaign for someone who was running for office'	-0.012 * (2.8)		
Respondent has 'ever contributed money to a candidate or a political party'	-0.006 (0.4)		
Number of 'projects or improvements' that the city has carried out 'that have helped you personally'			0.009 (0.5)
Constant	0.775	0.359	0.583
R^2(adjusted)	0.017	0.017	0.038

Table 5b.

Independent Variables	Dependent Variables — Favorable Responses Toward Spending on:	
	Veterans' bonus	Zoo
Education	-0.041* (17.0)	0.041* (12.3)
Yankee	-0.290 (11.9)	
Catholic	0.130* (13.6)	
'How much interest would you say you take in what goes on here in this neighborhood' (high score = high interest)	0.021 (1.3)	-0.014 (0.6)
Number of ethnic organizations of which member	-0.116* (5.1)	
Number of fraternal organizations of which member	0.112* (11.9)	
Number of labor organizations of which member	0.074* (4.2)	
Number of veterans' organizations of which member	0.093* (4.3)	0.093* (3.0)
Number of religious organizations of which member		-0.099* (7.2)
'Would you say that your religious upbringing was very religious, somewhat religious, or not so religious? ' (high = very religious)	-0.035* (6.7)	
Registered to vote	-0.039* (3.7)	
Number of political organizations of which member		0.214* (2.4)
Respondent has 'ever contributed money to a candidate or political party'		-0.028 (4.4)
Number of 'projects or improvements' that the city has carried out' that have helped you personally'		0.020* (2.1)
Constant	0.865	0.403
R^2 (adjusted)	0.152	0.067

Table 5c.

Independent Variables	Dependent Variables — Favorable Responses Toward Spending on:		
	Mayor/ council salaries	Police/ fire pensions	City hospital
Education	0.032* (8.6)		-0.012 (1.6)
Income	0.024* (6.2)	-0.018* (3.2)	
Length of residence in neighborhood		-0.017 (1.3)	
Believe in need for neighborhood groups		-0.019* (2.7)	-0.012 (1.7)
Head of household belongs to any club	-0.016* (3.3)		
Number of social organizations of which member	-0.035 (1.5)		
Number of fraternal organizations of which member		0.160* (16.8)	
Number of labor organizations of which member			0.072 (4.0)
Head of household is church member	0.019* (2.6)		
Number of three best friends of same religion			-0.109 (9.0)
Respondent has 'ever contributed money to a candidate or a political party'	-0.026* (5.4)		
Number of 'projects or improvements' that the city has carried out 'that have helped you personally'	0.027* (5.6)		0.018 (2.5)
Constant	0.110	0.583	0.798
R^2 (adjusted)	0.109	0.045	0.032

(3) Although only tenuous measures are available for our theoretical terms in the above equation, we do have data about participation of the respondent in a variety of ethnic, religious, and political activities. That several of these are independently significant in explaining support for spending suggests that they both help mediate the ethnic, religious, and income effects, and add a net increment.

Specifically, persons who have resided longer in the same neighborhood tend to favor more public school and welfare expenditures. However, other measures of neighborhood integration, such as a stated willingness to pay $5 to support a neighborhood organization, were either insignificant or negatively related to spending (for schools, welfare, and juvenile delinquency).

Turning to organizations, members of fraternal, labor, and veterans (but not ethnic) organizations favored the veterans' bonus.

Organizational membership taps a different dimension of activity from political participation, however, and effects on spending preferences are just the opposite. That is, persons registered to vote oppose the veterans' bonus, persons who contribute money to a political party oppose the zoo and mayor/council salary spending, and persons who report that they have 'ever helped in a political campaign' oppose public school spending.

These data suffer the normal weaknesses of most social science research, including less than perfect articulation between general concepts and specific indicators, and some degree of multicollinearity. However, the sample size is large enough to reduce intercorrelations among independent variables below 0.3, with the exception of education and income, as noted.[8] These specific results obviously provide only quite tentative support for our hypotheses as stated in the above equation.

NOTES

1. T. N. Clark, 'Catholics, Coalititions and Policy Outputs', in R.L. Lineberry and L.H. Masotti (eds.), Urban Problems and Public Policy (Lexington, Mass.: D.C. Death 1975), 65-78; T.N. Clark, 'The Irish Ethnic and the Spirit of Patronage', Ethnicity (1957), 2,305-359.

2. Cf. W. H. Riker and P. C. Ordeshook, An Introduction to Positive Political Theory (Englewood Cliffs, N.J.: Prentice-Hall 1973) for an overview.

3. See the paper by Tideman in this volume.

4. N. Frohlich, J. A. Oppenheimer, and O. P. Young, Political Leadership

and Collective Goods (Princeton, N.J.: Princeton Univ. Press 1971).

5. The sample and survey are discussed further in Clark, 'The Irish Ethic' and J. Q. Wilson and E. C. Banfield, 'Political Ethos Revisited', American Political Science Review, Vol. 65 (1971), 1048-1062.

6. Income data were obtained by using a card to elicit total family income using five income categories: under $3,000, $3,000-4,999, $5,000-9,999, $10,000-20,000, and $20,000 and over.

7. We recognize the statistical questions raised by using dichotomous dependent variables, as well as certain dummy variables as independent variables. Rather than apply log-linear or other procedures, we have chosen ordinary least squares regression to simplify analysis and communication.

8. Missing data for certain variables lowered the n for most regressions. The school regressions were computed using 924 cases; the others using 608.

Using Budget Pies to Reveal Preferences: Validation of a Survey Instrument

John P. McIver and
Elinor Ostrom

Why use a budget pie? A classic problem faces everyone who is curious about the perferences of at least one other individual: How can some simple questions lead us to accurately know about another's preferences? Husbands and wives sometimes face this problem when they are trying to find out where the other would like to vacation, what type of Christmas present the other would like, and so on. Some spouses try to hide their own preferences, deferring instead to the preferences of the other. Some married couples end up doing something neither partner prefers, both of them believing they are doing the thing the other most wants.

In the market-place, the entrepreneur faces the problem of ascertaining what his customers prefer. A misjudgement on his part will result in a loss of sales and profit. Continued difficulty in determining others' preferences will result in business failure. But a private entrepreneur does have some information to go on. As private consumers purchase goods, they reveal their preferences for certain goods at one price over other goods at other prices. In the short run, *quid pro quo* relationships reveal much about the preferences of buyers for the goods and services they are currently offered.

The authors are grateful to the Center for Studies of Metropolitan Problems of the National Institute of Mental Health for funding the data collection discussed in this paper (Grant 5 RO1 MH19911). Current support by the Research Applied to National Needs Division of the National Science Foundation (Grant GI-43949) and the Research Committee of Indiana University has enabled the authors to undertake this analysis. The authors also wish to thank Terry N. Clark, whose footnoting of this research and continuous pressing to get findings into print forced us to write this paper. The authors also thank Jnana Hodson, Roger B. Parks, Vernon L. Greene, and the staff of the Workshop in Political Theory and Policy Analysis for their help in the editing and preparation of this paper.

Public entrepreneurs, on the other hand, obtain little information from daily exchanges about the current preferences of their 'consumers'. Because the goods and services most public entrepreneurs produce are made available to a defined clientele without any *quid pro quo* relationships, no record of the satisfaction that clientele receive from these goods and services is produced. This is the case with public goods such as police. For instance, the residents of a city have no short-run choice about the type and level of police services offered to them by their local police department. Citizens cannot select from several competitive producers. They cannot 'purchase' the mix of police services that best satisfies their own preferences at a particular set of prices. Services are delivered with very little information coming back to the community's elected officials or to those responsible for the police department's day-to-day operations.

What information is obtained from citizens is difficult to interpret. Because the payment for such goods is separated from the delivery of the goods, most individuals want 'more' rather than 'less' of any public goods, as long as they place a positive value on the goods. (Most individuals also prefer more of most private goods to less of them, but the necessity of paying for any amount of such goods forces them to reveal their preferences for goods in light of their costs.)

Social scientists interested in the relationship between citizens' preferences for public goods and the responsiveness of public officials to these preferences must also face this problem. If survey questions simply ask whether citizens would like more or less of a particular service – without any budgetary constraint – respondents will obviously tend to respond that they want more.

Clark addresses this question in an important paper, 'Can You Cut a Budget Pie?'[1] He identifies several modes of asking survey questions that encourage respondents to truthfully reveal their preferences for different levels of public goods in light of a bugetary constraint. Among the modes he presents is the 'budget pie'. A respondent is presented a circle and asked to 'cut the pie' into segments that represent the respondent's preferred allocations for providing some public good or service. Clark critically assesses the strengths and weaknesses of this format and concludes:

> The budget pie is an appealing format in generating, tentatively, more information about values than many alternative instruments. But if in principle it can achieve many ideals, precise results depend on several structural supports that can give way: conversion of money to performance, a direct relation of money to utility, honest preference revelations, etc.

These suggest that for certain populations, under certain conditions, the budget pie can be ideal, but for others it is grossly improper.[2]

An early version of Clark's paper was received by members of the Workshop in Political Theory and Policy Analysis at Indiana University when we were in the preliminary stages of designing a comparative study of police performance in the St Louis metroplitan area. Since little empirical experimentation with this type of question format had been undertaken, it seemed reasonable to try to use this type of question in our attempt to assess citizen preferences for different types of police activities. The major research questions to be addressed involved citizens' experiences with the police and evaluations of police performance. The relationship of citizens *preferences* for a public service to their *evaluation* of that service was a corollary research question.

THE SURVEY INSTRUMENT

Several budget pie question formats could have been used. One could have shown a citizen different levels of general police services, each with its own cost figure. The page handed to a respondent would have had several differently-sized budget pies on it, each with a description of an associated service-level. In the design stage, we experimented with such a format. Phrasing police service levels in a simple manner turns out to be very difficult. Early pre-tests eliminated this format from further consideration in our study. Where researchers have sufficient time to test different descriptions of service levels, we would urge continued experimentation with this type of format. A second format could involve asking a respondent to allocate the entire city budget to several key services including police.

We were, however, particularly interested in preferences for different types of police activities and decided to use a third type of format. We asked respondents to allocate the total police budget for their jurisdiction among three types of police activities: patrolling, detective work, and administration.[3] The sheet we handed each respondent is reproduced as Figure 1. The interviewer read the instructions to the respondent; these were written at the top of the page. When respondents seemed to be having difficulty, the interviewer suggested they 'think of it as cutting a pie into three pieces. The size of the piece will indicate how much of the budget you think should be spent on the activity.'

Next, we would like you to assume that you are in charge of deciding how the police should spend their budget dollar on the following three activities:

A.   Patrolling (crime prevention)
B.   Detective work (criminal investigation)
C.   Administration

Indicate by dividing this budget dollar into three separate sections the way *you* would like to see the police spend their budget. You can identify each section by writing the letter A, B, or C in the corresponding sector.

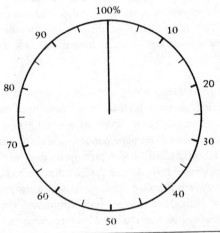

Figure 1. Budget Pie Question Format Utilized in this Study

This question asked respondents their preferences in regard to how their local police departments should allocate resources among different police activities. We also gathered data in each of the twenty-nine departments serving our sample neighborhoods regarding the allocation of police resources to patrolling, criminal investigation, and administrative activities. An original hypothesis was that citizens living in jurisdictions where police department allocations were rather congruent to the residents' own preferences would evaluate police service levels more affirmatively than would citizens living in other jurisdictions.

Many assumptions about respondent information levels underlie the use of this question format. Our assumption is that citizens have preferences for the ways a public agency uses its resources to support varying levels of activities. Such a preference would, in essence, be a 'preferred production function' for the service involved. To have such a preference, citizens would have to have some basic conception of how different combinations of activities might affect the output of police services.

THE DATA

The data were collected during the spring of 1972 in the St Louis metropolitan area. Utilizing a 'multi-strata, most similar systems' research design described in detail elsewhere,[4] a series of in-person, mail, and telephone instruments were administered to approximately 4,000 citizens. The budget pie question however, was included in only the in-person instrument. The in-person instrument was administered to 1,791 respondents. Of this total, 83 percent or 1,479 respondents filled out the budget pie sheet. Of these respondents, 94 percent (1,401) filled out the form correctly. (A form was filled out correctly if the respondent allocated the full budget pie to the three services without any overlap among expenditure allocations.)

The budget pie was administered during the first third of an interview requiring about forty minutes. We found that this question format was more difficult to administer than were many of the more typical question formats used in the remainder of the instrument. The difficulty was related to the socio-economic status (SES) of respondents. In middle-class neighborhoods, where respondents were primarily homeowners used to receiving tax bills and to examining city budget allocations represented as slices of a budget pie in local newspapers, a higher proportion of the respondents completed the budget pie correctly. The most difficult problems in administering the format came in lower SES neighborhoods, where respondents were less familiar with thinking about budgetary allocation as represented by a budget pie. In the two lowest SES neighborhoods, budget pies were completed by only 70 percent of respondents and of those completed, only 75 percent were completed correctly, in contrast to the 95 percent completed correctly in the higher SES neighborhoods.

From this field experience, we would not recommend using this budget pie format where considerable variation in the SES of respondents would be present. Given our St Louis research design, the variation we encountered is less than would be the case in a random sample of an entire metropolitan area. On the other hand, if future researchers were interested in a sample of middle- or upper-class homeowners, the difficulty would be minimal.

VALIDATION OF A BUDGET PIE

How valid is the budget pie as a method for measuring preferences for police services? At least one researcher, Douglas Scott, would question its use as the *only* measure of preference:

> ... while the use of budget pies provide a superior (e.g., ratio) level of measurement per se, it is not immediately clear in what context citizen preferences are being made ... (M)ultiple measurements of preferences would provide more insight into the psychological meaning as well as the substantive meaning of choice.[5]

Scott would require that we validate our budget pie using multiple methods to measure preferences.

Numerous validation criteria have been proposed by social scientists and policy researchers in many fields.[6] Two of the most important types of validity discussed are convergent validity and construct validity.

Convergent Validity

Convergent validity may be thought of as confirmation by independent measurement procedures. In our effort to establish the validity of the budget pie, we examined the convergence of two methods of measuring preferences for police services.[7]

Another technique used in the survey to ascertain specific aspects of citizen preferences for police services in their neighborhood was an open-ended question asking respondents to indicate needed improvements in law enforcement in their neighborhood.[8] Coding of this question yielded a number of categories that should be related, if our measures are to be convergently valid, to respondents' preferences for police budget allocations as revealed by their response to the budget pie. In other words, if our preference measures are valid, one would expect respondents to express similar preferences on both types of questions. The correlations between preferences for expenditures as expressed by 'cutting the pie' and preferences for improvements in police services are reported in Table 1.[9]

Although small, the correlations support the conclusion that our preference measures are consistent, a minimum requirement for convergent validity. For example, the validity of a preference for patrol as expressed by cutting the budget pie is enhanced by the correspondence of this preference with a preference for general patrol and foot patrol measured by an alternate method. Furthermore, support for improved control of narcotics is seen to correlate with

Table 1. Yule's *Q* Correlations Between Preferences for Budgetary Allocations for Police Activities (the Budget Pie) and Preferences for Improvements in Police Services (Open-ended Question)

Use Federal or State Grants for Improvements in:	Amount of Budget Pie Allocated to:		
	Patrol	Criminal Investigation	Administration
General Patrol	0.20*	0.00	-0.11
Foot Patrol	0.39	-0.40	-0.05
Police-community relations	0.27	-0.08	-0.11
Additional officers	0.10	-0.10	0.01
Narcotics control	-0.19	0.19	0.21
Juvenile problems	-0.02	0.02	0.17
Support facilities (new equipment — additional improved stations)	-0.18	0.08	0.10
No improvements	-0.15	-0.02	0.14

*Significant at the 0.05 level; Note: $N = 1307$.

preferences for increased expenditure for criminal investigation and supportive service activities. Not to overstate the support for convergent validity, we note that a substantial lack of co-variation does exist between indicators. We do not, therefore, have strong evidence for the convergent validity of these measures. Campbell and Fiske have outlined several alternate situations that might be in effect if little (or no) convergent validity is found:

(1) Neither of our methods is adequate for measuring preferences;
(2) One of the methods does not measure preference; or
(3) The construct 'preference' is not a functional unity, or the responses evoked are specific to non-construct attributes of each measure of preference.[10]

The second or third situation may hold in this instance. It is more likely that the third case is in effect; we are actually asking for separate aspects of citizen preferences when dealing with preferences for expenditure allocations and improvements based on an expansion of the pie. We feel that the small but consistent coefficient should not lead to rejection of the budget pie as a measure of preference, but should lead to further development of the technique. We used one of the most simplified formats for the budget pie. Through our experience in the field, analyses using this pie, and work by other researchers, the question format may be improved.

Construct Validity

For scientific purposes, the most important characteristic of a measuring instrument or test is its *construct validity*. This type of the validity of a measure, or test, of a construct is the degree to which it ties into a network of related concepts. As Runkel and McGrath indicate, 'It requires the prior establishment of a network of relations among a set of operational measures which are presumed to be valid measures of certain related conceptual properties.'[11] Construct validation has three parts, according to Cronbach.[12] First, one suggests what construct possibly accounts for test performance. In our case, we are suggesting that it is preference for police budget allocations which underlies response to the budget pie question. Second, hypotheses are derived from the theory involving the construct. Finally, the hypotheses are tested empirically. Kerlinger indicates:

> ...construct validation and empirical scientific inquiry are closely allied. It is not simply a question of validating a test. One must try to validate the theory behind the test.[13]

An inherent difficulty in construct validation is that the construct is not directly measurable. The test-construct correlation cannot, therefore, be computed. Theory plays an important role here. It is possible to compute the correlations between the test by indicating the construct under investigation and tests that, according to theory, should correlate with the construct *and* other tests that should not correlate with it. Three predicated relationships, therefore, are possible: positive, negative, or zero. (For this reason, construct validity is alternatively termed 'differential prediction' or 'differential validity'.) An operational measure is valid if it relates in the way the construct 'should' relate to measures of other related constructs.[14]

As part of the process of construct validation, we can pose a simplified linkage model in the form of two questions:

(1) Are public agencies responsive to citizen preferences?
(2) Are citizens more satisfied with services received when public agencies are more responsive to their preferences?

These are certainly not the only questions relevant to agency/ constituent interaction. For instance, to avoid complicating the discussion, we ignore the effects of information. Clearly the negligible transmission of information from public agencies to consumers and of demands from consumers to public agencies has a considerable negative effect in the responsive conversion of demands. As a result, support for the model and therefore the validity of the measurements will be underestimated.

Lacking time-series data necessary to make causal inference about the process, one might examine whether agencies allocate their budgets in a congruent manner with the preferences of citizens living in their jurisdictions in order to make some inference about agency responsiveness.[15] This model of agency 'responsiveness' could be posed in the following manner:

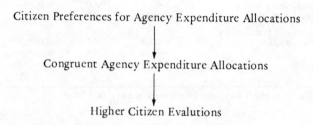

Citizen Preferences for Agency Expenditure Allocations

Congruent Agency Expenditure Allocations

Higher Citizen Evalutions

Adequate examination of this model hinges on the answers to a number of methodological questions. The most important of these deal with measurement validity. Assuming that a citizen has a preference for specific expenditure allocations, our primary interest is in validating an instrument to measure that preference.

Citizen Preferences and Expenditure Allocations

According to our postulated model, citizen preferences for agency expenditure allocations should correlate with actual expenditure allocation patterns.[16] But Table 2 shows little association between preferences for expenditure allocation by respondents and actual expenditure allocation by their own police departments. Having assumed the correctness of the model to test the validity of our measurement instrument, we must initially question whether the budget pie adequately measures preferences. We have not yet fulfilled the requirements to test construct validity — to examine 'preferences' within a network of theoretical relationships with other constructs.

As a next step, therefore, we examine the relationship between preferences for expenditure allocations and actual expenditure allocations using three controls: neighborhood SES, neighborhood

Table 2. Pearson Product-Moment Correlations Between Preferences for Budgetary Allocations for Police Activities (Budget Pie) and Police Agency Expenditures, with Controls

| | Police Activities | | | |
	Patrol	Criminal Investigation	Administration	N
Zero Order	0.00	0.05*	-0.05*	1307
Neighborhood SES				
Upper Middle	0.10*	0.12*	0.04	313
Lower Middle	-0.02	0.04	-0.07*	966
Neighborhood Racial Composition				
Predominantly White	-0.01	0.00	0.00	994
At Least 30% Black	0.02	0.15*	-0.17 i	313
Population of Jurisdiction				
80,000-623,000	0.04	0.01	-0.09*	428
16,000-80,000	0.01	0.04	-0.07*	624
0.16,000	-0.02	0.03	0.00	255

*Significant at the 0.05 level; + Significant at the 0.001 level

Note: Neighborhood SES is operationalized as median value of owner occupied housing and median contract rent.

racial composition, and jurisdiction population size. Adding these controls to the model enables us to incorporate Clark's assertion, 'For certain population, under certain conditions, the budget pie can be ideal, but for others it is grossly improper.'[17]

We do find stronger, significant correlations between preferences for expenditure allocations and actual expenditure allocations in higher SES neighborhoods than in lower SES neighborhoods. One might expect residents of higher SES neighborhoods to be more effective in articulating their preferences for the allocation of resources than are residents elsewhere. Alternatively, we might also expect respondents in higher SES neighborhoods to have more information about how resources are normally allocated in police departments. In particular, none of the police departments included in the study allocate less than 28 percent to administration while over 70 percent of our respondents allocated less than 28 percent to administration. Since respondents in high SES neighborhoods tended to allocate a higher proportion of their budget pies to administration, the correlation between preferences and evaluation could be expected to be somewhat higher. This relationship is also consistent with our field experience, where respondents in higher SES neighborhoods found the budget pie format much easier than did respondents in other neighborhoods.

The significant correlations between preferences and expenditures in neighborhoods having a high proprotion of black residents are puzzling. These correlations could be spurious. In general, these neighborhoods are characterized by a higher crime rate. The higher incidence of victimization may be leading both police agencies and residents in these neighborhoods to increase allocations to criminal investigation. Thus, it is not possible to ascertain from this analysis whether the positive association between citizen preferences for allocations to criminal investigation and actual budgetary allocations to this activity is caused by a third variable. The negative relationship between respondents' preferences for allocation to administration and actual allocation to this activity might result from two factors. As respondents increased their allocation to criminal investigation, they were more likely to cut back on administration than on patrol. Agencies, on the other hand, are more likely to cut back on patrol than on administration when allocating more to criminal investigation.

Controls for jurisdiction size did not affect the zero-order relationship significantly, although we had hypothesized that population size — as a surrogate for linkages between citizens and officials — might have a significant effect on the way demands for services are translated into activities.

Congruence and Evaluation

The second linkage in our model is the relationship between citizen *evaluations* of police services and the *congruence* of police agency expenditures with citizen preferences. Congruence is operationalized as the difference between the proportion of the pie allocated to an activity by a respondent and the percentage of the police agency expenditures on that activity. Total congruence is simply the sum of the congruence between preferences and expenditures for all three police activities. For our evaluation variable, we used a scale of seven items contained in the same survey instrument.[18] Our original hypothesis was that high congruence is postively related to high evaluation of police services. The data support this hypothesis for all police activities.

The relationship between evaluation of police services and congruence of police agency expenditures with citizen preferences for agency allocations may be complicated by the meaning an 'expressed preference' has for each respondent. A high preference for expenditure allocation to a police activity may mean one of two things. A respondent may value and evaluate that activity highly. The respondent may, therefore, express a preference for greater allocation of budgetary resources to that activity over others. *Or,* a respondent may consider the current level of activity insufficient and thus prefer more spending for that activity. If both types of respondents are included in the sample, analysis based on aggregation of these two types of individuals may lead to the conclusion that no correlation betweeen preferences and evaluations exists even though, in fact, positive and negative correlations may exist for subgroups of the population.

Consequently, we examined this linkage by using several controls that may help to isolate groups of individuals who tend to have similar outlooks on police services. (We cannot actually control for the different mental operations by which preferences and evaluations are related. Thus, we have chosen our groups on the basis of known differences in their perceptions toward and experiences with crime and the police.) Black respondents, for example, have been found in a number of surveys to be more critical of police services in their neighborhoods than are white respondents.[19] Respondents who have been victimized are frequently more critical, while those who have been assisted are more supportive of their local police.[20] Renters, who frequently experience more burglaries, can be expected to be more critical than are homeowners.[21] On the other hand, homeowners, having a greater investment in the community, may be more critical of the services they receive. But homeowners also tend to remain in the same neighborhood for longer periods of time which frequently results in a more positive evaluation of the neighborhood and the services they receive from their local government.

Clark distinguishes two groups as possibly reacting strategically to the use of the budget pies to ascertain preferences for police expenditures. He suggests that strategic responses may be more frequent (1) when the respondent is well-informed about the services received|; and (2) where the 'free-rider' problem will be most obvious (that is, in smaller groups where the respondent may be more certain that others will vote to provide adequate levels of service and where the respondent has a greater ability to influence outcomes with a single vote).[22] A final set of controls, based on a classification of the strategic conditions under which the budget pie question was answered, will be examined. A four-category typology of these situations is found in Figure 2. We used the four-category classification of respondents derived from Figure 2 as control variables in Table 6.[23]

Figure 2. A four-fold Typology of Strategic-Response Situations

Information Levels

	High	**Low**
Large	Group 1	Group 2 **Least Strategic**
Small	Group 3 **Most Strategic**	Group 4

Jurisdiction Size (row label on left)

Our complete set of control variables can be placed into four broad categories: respondent perceptions of police services, respondent experience with crime and the police, personal characteristics of the respondent, and respondent strategies.

Respondent perceptions of current service levels can affect both their evaluations of police as well as their preferences for services. Indeed, perceptions of service may well cause greater variation in these variables, more than the quantity and quality of actual services received. Controls for perceived police response time, estimated number of neighborhood patrols, and perceived mistreatment of others by the police are incorporated into our basic model in Table 3. While the correlations are not all significant, due to small n's, the associations between congruence and evaluation increase for a number of the control groups. This is especially true when knowledge of police mistreatment of other members of the community is controlled.

Citizen-police interaction and experience with crime also has an affect as a supressor variable. In Table 4 we report our linkage relationships controlling for whether or not a respondent has been assisted

Table 3. Pearson Product-Moment Correlations Between Evaluations
of Police Services and the Congruence of Citizen Preferences
for Budgetary Allocations and Agency Expenditures, with
Controls

| | Police Activities | | | | |
	Total	Patrol	Criminal Investigation	Administration	N
Zero Order	0.10	0.08	0.05	0.05	1236
Respondent knows another mistreated by police					
Yes	0.09	0.14*	-0.11	0.11	169
No	0.10+	0.04	0.08*	0.06*	1040
Respondent's perceived police response rate					
Rapidly	0.07*	0.10*	0.03	-0.01	648
Slowly	0.07	0.02	0.03	0.07	456
Respondent's perceived number of police patrols in neighborhood					
At least three times/shift	0.07*	0.07*	0.06	0.00	626
Two times/shift-once a week	0.09*	0.09*	0.05	0.02	420

*Significant at the 0.05 level; + Significant at the 0.001 level.

Note: The relationships between total congruence and evaluation are found
under 'Total'.

or stopped by a police officer and whether or not the respondent
has been the victim of a crime. Together with perceived mistreatment,
each of these controls may increase the salience of police services
received. Each of these controls was, therefore, expected to have an
effect on the relationship between congruence and evaluation.
Comparison of the correlations with and without controls supports
this hypothesis.

A number of respondent characteristics are significant as factors
affecting responses to the budget pie question format or the
relationship between congruence and evaluation. In Table 5, the basic
model is examined with controls for the respondents' race, sex,
occupation, education, years in residence in the neighborhood, and
home ownership status.

Home ownership proved to be the most influential control variable
in our analysis. The product-moment correlation between total
congruence and evaluation is 0.21 greater for renters than for home

Table 4. Pearson Product-Moment Correlations Between Evaluations of Police Services and the Congruence of Citizen Preferences for Budgetary Allocations and Agency Expenditures, with Controls

| | Police Activities | | | | |
	Total	Patrol	Criminal Investigation	Administration	N
Zero Order	0.10+	0.08*	0.05*	0.05*	1236
Respondent stopped by police					
Yes	0.19+	0.10*	0.16+	0.09*	355
No	0.06*	0.07*	0.00	0.03	878
Respondent has been victimized					
Yes	0.13*	0.09	0.10*	0.06	305
No	0.09*	0.08*	0.04	0.04	931
Respondent has been assisted by police					
Yes	0.09	-0.04	0.15*	0.05	283
No	0.10+	0.11 +	0.02	0.05	953

*Significant at the 0.05 level; + Significant at the 0.001 level.

Note: The relationships between total congruence and evaluation are found under 'Total'.

buyers. Respondents' occupation, education, sex, number of years in residence, and race also affect the zero-order relationships found in Table 5, but not to the extent home ownership does.

Table 6 presents the relationship between congruence and evaluation of police service levels under our controls for strategic conditions. If Clark is correct about the effects of group size, we would focus on groups 1 and 2 in comparison to groups 3 and 4. It is quite evident that there are significant differences in the applicability of the budget pie to certain groups. The correlations between total congruence and evaluation for groups 1 and 2 is 0.15 and 0.16, significant to at least the 0.01 level, while this relationship is not significant in the 'small' groups (although the association is positive, the direction predicted by our responsive agency model). The amount of information on police services does not affect the relationship to the extent that size does. While informed respondents do have incentives to act strategically, it is likely that the uncertainty concerning the use of and authority behind the survey will minimize these

Table 5. Pearson Product-Moment Correlations Between Evaluations of Police Services and the Congruence of Citizen Preferences for Budgetary Allocations and Agency Expenditures, with Controls

| | Police Activities | | | | |
	Total	Patrol	Criminal Investigation	Administration	N
Zero Order	0.10+	0.08*	0.05*	0.05*	1236
Respondent's Occupation					
Hi	0.06	0.07*	0.13*	-0.06	513
Lo	0.13+	0.08*	0.00	0.12*	697
Respondent's education					
H.S. or Less	0.12+	0.08*	0.02	0.09*	798
More than H.S.	0.10*	0.06	0.14*	0.00	432
Respondent's home ownership status					
Renter	0.28+	0.18*	0.21*	0.13	142
Owner	0.08*	0.06*	0.03	0.04	1084
Sex					
Male	0.12+	0.07*	0.11*	0.05	589
Female	0.08*	0.08*	0.00	0.04	645
Race					
Black	0.05	-0.06	0.03	0.11*	240
White	0.11+	0.09*	0.09*	0.02	992
Years in Neighborhood					
10 years or less	0.10*	0.08*	0.01	0.08*	707
More than 10 years	0.08*	0.05	0.12*	0.00	529

*Significant at the 0.05 level; + Significant at the 0.001 level.

Note: The relationships between total congruence and evaluation are found under 'Total'.

incentives.

Clearly, there are many complex interactions between the control variables that we cannot deal with in this paper. In attempting to examine the validity of the budget pie, we have reported as much information as we could to permit others to make their own judgements about the usefulness of this instrument. We have not offered hypotheses on the full range of the relationships between preference, expenditures, congruence and evaluation for patrol criminal investigation and administrative activities due to *lack* of theory and research to guide hypothesis testing regarding the specific activities of the police. We hope we have substantiated this procedure at least to

Table 6: Pearson Product-Moment Correlations Between Evaluations of Police Services and the Congruence of Citizen Preferences for Budgetary Allocations and Agency Expenditures, with Controls

| | Police Activities | | | | |
	Total	Patrol	Criminal Investigation	Administration	N
Zero Order	0.10+	0.08*	0.05*	0.05*	1236
Respondent's strategic context					
Group 1	0.16*	0.11*	0.01	0.17+	326
Group 2	0.15*	0.08	0.07	0.12*	249
Group 3	0.07	0.06	0.02	0.03	367
Group 4	0.10	0.04	0.11	0.02	258

*Significant at the 0.05 level; + Significant at the 0.001 level.

Note: The realtionships between total congruence and evaluation are found under 'Total'.

the point of stimulating further exploratory work dealing with the problem of obtaining 'constrained' preferences in citizen surveys.

CONCLUSION

In exploring the validity of using budget pie formats to encourage respondents to reveal their preferences for public goods and services, we can make several general observations. From our fieldwork, we know this format is limited to applications in surveys where the sample frame is predominantly middle- and upper-SES respondents. Given the difficulties in both administering and gaining completed forms among lower-SES respondents, any survey using this format across a broad SES range would find the response rate among lower-SES respondents considerably reduced.

From our analysis, some weak support exists for convergent validity of the budget pie format. Respondents who wished to see additional patrol and foot patrol in their local police, for example, were also more likely to allocate a larger proportion of the budget pie to patrol than were respondents who wished to see improvements for non-patrol related activities. But although the relationships on similar but different measures were significant, they were not strong.

With regard to construct validity, we examined the posited linkages among three variables: citizen preferences for budgetary allocations to police activities; congruence of agency budgetary allocations for those activities; and citizen evaluations. The first linkage was between citizen preferences for budgetary allocations to police activities and actual budgetary allocations to patrol, criminal investigation, and administration. The linkage was not supported by the data, except for residents of high-SES neighborhoods and a mixed and puzzling pattern for residents living in neighborhoods having a higher proportion of black residents.

But where congruence did exist, we found a consistent and positive, but weak, relationship between respondents served by police agencies with a budgetary allocation relatively congruent to the respondent's own evaluations of the local police. Introducing control variables slightly increased this relationship for at least one subgroup in each analysis.

Our analysis of the question of validity would lead us to have some increased confidence in the use of budget pies for obtaining accurate reflections of citizen preferences, but also to have continued modesty in the level of success achieved through its use. We hope other researchers will experiment with other formats of the budget pie. We believe that our own format, which required a knowledge of the production process, probably assumed too high a level of information. Budget pies focusing on the amount allocated to services rather than to activities, while they may be more difficult to design, might be more successful.

APPENDIX

A major problem with the use of Likert scales and related procedures for measuring preferences is that a respondent may be encouraged to overstate his or her preferences for public goods and services. It is therefore necessary to impose constraints in order to get a more accurate picture of respondents preferences. We have suggested that the budget pie be considered as one method which imposes the necessary constraints. But what effect will these constraints have on analyses using data collected by the budget pie method? Two important results, dealing with the relationship between sections of the pie and the correlations between each section of the pie and a criterion variable, will be illustrated below.

Analytically, the budget pie may be considered as a sum of proportions totalling 1.0,

$$1.0 = \sum_{i=1}^{n} P_i \qquad (1)$$

such that any section of the pie, j, is equal to 1 minus the sum of all other sections of the pie

$$j = 1.0 - \sum_{c=1}^{n} P_i \qquad i \neq j. \qquad (2)$$

With this definition of the budget pie and each section, we can explore the effects of the data collection method on analysis. In order to simplify discussion, we will deal with a three-section pie.

Correlations betweeen segments of a budget pie
Given a budget pie with three sections, x, y and z, the correlation between any two sections is equal to the covariance of these sections divided by the product of the standard deviations of those segments. For example,

$$\rho_{x,y} = \frac{\sigma_{xy}}{\sigma_x \sigma_y} \qquad (3)$$

Defining y in terms of the other segments of the pie, that is,
$y = 1.0 - x - z$

$$\rho_{x,1-x-z} = \frac{\sigma_{x,1-x-z}}{\sigma_x \sigma_{1-x-z}} \qquad (4)$$

This expression may be reduced as follows:

$$\rho_{x,1-x-z} = \frac{\sigma_{1x} - \sigma_x^2 - \sigma_{xz}}{\sigma_x \left(\sigma_1^2 + \sigma_x^2 + \sigma_z^2 + 2\sigma_{xz} + 2\sigma_{1x} + 2\sigma_{1z} \right)^{1/2}}$$

$$= \frac{-(\sigma_x^2 + \sigma_{xz})}{\sigma_x \left(\sigma_x^2 + 2\sigma_{xz} + \sigma_z^2 \right)^{1/2}}$$

$$= \frac{-\left(\sigma_x^2 + \rho_{xz}\sigma_x\sigma_z \right)}{\sigma_x \left(\sigma_x + \sigma_z \right)}$$

$$= \frac{-\left(\sigma_x + \rho_{xz}\sigma_z \right)}{\sigma_x + \sigma_z} \tag{5}$$

Assuming homoscedasticity (equal variances), i.e. $\sigma_x - \sigma_z$

$$\frac{-(1 + \rho_{xz})}{2} \tag{6}$$

This expression will be negative unless $\rho_{xz} = -1.0$. The researcher using a budget pie should be aware of multicollinearity problems in correlating variables (sections of the pie) which are part of the same linear composite with a dependent variable. This result will be true for any technique which attempts to constrain preferences. The correlations between sections of the pie will be reduced but not completely eliminated by increasing the number of choices over which preferences may be expressed.

Correlations of all budget pie sections with an independent variable (A)

By definition

$$\rho_{x,a} = \frac{\sigma_{xa}}{\sigma_x \sigma_a} \tag{7}$$

Defining x in terms of the rest of the pie, that is, $x = 1 - y - z$,

$$\rho_{1-y-z, a} = \frac{\sigma_{1-y-z, a}}{\sigma_{1-y-z}\ \sigma_a}$$

$$= \frac{-\sigma_{ay} - \sigma_{az}}{(\sigma_y + \sigma_z)\ \sigma_a}$$

$$= \frac{\rho_{ay}\sigma_a\sigma_y + \rho_{az}\sigma_a\sigma_z}{\sigma_a\sigma_y + \sigma_a\sigma_z} \tag{8}$$

Assuming homoscedasticity of pie sections, i.e. $\sigma_y = \sigma_z = \sigma$

$$= \frac{-2\sigma_a\sigma\ (\rho_{ay} + \rho_{az})}{2\sigma_a\sigma}$$

$$= -\rho_{ay} - \rho_{az}. \tag{9}$$

Summing the correlations of all sections of the pie with A.

$$\rho_{ax} + \rho_{ay} + \rho_{az} = (-\rho_{ay} - \rho_{az}) + \rho_{ay} + \rho_{az}$$

$$= 0.0. \tag{10}$$

We have just demonstrated that the tradeoff of services required by the budget pie will necessarily be present in data anlaysis. Again, this is because our measure of preference is a linear composite. Furthermore, this appendix should prevent overstating the case for construct validity in using constrained preferences: a predicted positive correlation between one section of the pie and a criterion variable may necessitate a negative correlation between the criterion variable and another section of the pie. Ordinarily this would be counted as two tests of the validity of the pie. However, it is not the case that one is making two independent tests due to the nature of the measuring instrument.

NOTES

1. T. N. Clark, 'Can You Cut a Budget Pie?', Policy and Politics, Vol. 3 (1974), 3-31.

2. Ibid., 26.

3. We consider activities an intermediate step toward the production of services or output. Thus police do patrol, investigate crimes, and administrative activities which themselves contribute in some way to the production of police services, such as rapidly responding to calls, solving crimes, and assisting citizens. Our budget pie, therefore, required the respondent to indicate a preference for intermediate activity levels in the production process.

4. E. Ostrom, R. B. Parks and D. C. Smith, 'A Multi-Strata, Similar Systems Design for Measuring Police Performance', paper presented at the Midwest Political Science Association Annual Meetings, May 1973. D. C. Smith and E. Ostrom, 'The Effects of Training and Education on Police Attitudes and Performance: A Preliminary Analysis', in H. Jacob (ed.), The Potential for Reform of Crimincal Justice 3, Sage Criminal Justice System Annuals (Beverly Hills and London: Sage 1974), 45-81.

5. D. Scott, 'Citizen Evaluations of Local Government Services: Some Measurement Questions', Report 192 of the Institute of Government and Public Affairs, University of California, Los Angeles (November 1974), 8.

6. See, for example F. Kerlinger, Foundations of Behavioral Research (New York: Holt, Rinehart and Winston 1964), chapter 25 for a summary of the various types of validity and the American Psychological Association et al., 'Technical Recommendations for Psychological Tests and Diagnostic Techniques', Psychological Bulletin, Vol. 51 (1954), Supplement, for a more extensive statement of validity criteria. Two important papers should also be noted: L. Cronbach and P. Meehl 'Construct Validity in Psychological Tests', Psychological Bulletin , Vol. 52 (1955), 281-302 and D. T. Campbell and D. W. Fiske, 'Convergent and Discriminant Validation by the Multitrait-Multimethod Matrix', Psychological Bulletin, Vol. 56 (1959), 81-105.

7. Independence of measurement procedures is the major difference between validity and reliability. It is a requirement of both that agreement between measures be demonstrated. But reliability is the agreement between two attempts to measure the same construct through maximally 'similar' methods, while validity is demonstrated by agreement between measurements using maximally 'different' methods.

8. This question reads: 'Suppose money were available from a federal or state grant for one major improvement in law enforcement in your neighborhood. What one improvement should be made?' This question is another attempt to constrain preferences, in this case by limiting improvements to the service the respondent sees as most necessary in that it is his first reaction.

9. All correlations reported in this paper with the exception of Table 1 are Pearson product-moment correlations. We use this statistic to gain maximum information from our interval level measurements of preferences, expenditures,

and evaluations. In Table 1, responses to the open-ended question regarding referred improvements in services can, at best, be regarded as ordinal. In comparing budget pie responses to the responses to the open ended question, the 'pie' response for each of the three police activities is dichotomized into 'high' and 'low' categories. Yule's Q is the measure of association used to analyze the relationship between the two measures of preferences. An excellent discussion of the use of Q can be found in J. A. Davis, Elementary Survey Analysis (Englewood: Prentice Hall 1971).

10. Campbell and Fiske, op. cit., 104.

11. P. J. Runkel and J. E. McGrath, Research on Human Behavior: A Systematic Guide to Method (New York: Holt, Rinehart and Winston 1972), 162.

12. L. Cronbach, Essentials of Scientific Testing, 2nd edn., (New York: Harper and Row 1960), 121.

13. F. Kerlinger, op. cit., 449.

14. The test may correlate with another test with which theory suggests it should not correlate. This may be the case when the tests correlate 'spuriously', that is, when they both measure in part something other than the construct under investigation. This may be due to another construct, systematic errors of measurement, etc. The researcher's responsibility is to investigate such instances in demonstrating construct validity.

15. Congruence as a measure of 'responsiveness' has been utilized most recently by S. B. Hansen, 'Participation, Political Structure, and Concurrence', American Political Science Review, Vol. 69 (1975), 1181-1199. Similar measures, though dealing more obliquely with the problem by correlating attitudes rather than preferences. S. Verba and N. H. Nie, Participation in America: Political Democracy and Social Equality (New York: Harper and Row 1972) and W. E. Miller and D. Stokes, 'Constituency Influence in Congress', American Political Science Review, Vol. 57 (1963), 45-56. While we have noted the problems with a measure of congruence as responsiveness, it is probably the closest surrogate available in cross-sectional survey research.

16. Expenditures by police agencies on the three police activities in the 44 neighborhoods in our sample are estimated on the basis of departmental structure and neighborhood demands (percentage of communications dispatches). These formulae were derived by Roger B. Parks and may be found in E. Ostrom. W. H. Baugh, R. Guarasci, R. B. Parks and G. P. Whitaker, Community Organization and the Provision of Police Services (Beverly Hills: Sage 1973) 53-57.

17. T. N. Clark, op. cit., 26.

18. A seven-item scale is used as a measure of respondents' 'evaluations' of his or her local police department. These seven items included a general rating of police services, a rating of police community relations, evaluations of police honesty, courtesy and professionalism, a question regarding the respondent's confidence in the police, and support for a rejection of a statement that police treat everyone equally. The mean inter-item correlation is 0.46 and the reliability coefficient for the scale ('alpha') is 0.85. A number of respondents did not respond to several of the items which comprised the scale. As a result, for analyses using the evaluation scale, $N = 1236$.

19. J. D. Aberbach and J. L. Walker, 'The Attitudes of Blacks and Whites Toward City Services: Implications for Public Policy', in J. P. Crecine (ed.),

Financing the Metropolis Vol. 4 Urban Affairs Annual Reviews (Beverly Hills: Sage 1970), 519-538.

20. E. Ostrom, W. H. Baugh, R. Guarasci, R. B. Parks and G. P. Whitaker, op. cit. 41.

21. A. L. Schneider, 'Crime and Victimization in Portland: Analysis of Trends, 1971-1974', Oregon Research Institute (1975), 42.

22. T. N. Clark, op. cit., 21.

23. Respondent's information level is operationalized as the ability to answer a series of thirteen questions on the local police services the respondent receives.

24. T. N. Clark, op. cit., 13.

Measures of Citizen Evaluation of Local Government Services

Douglas Scott

INTRODUCTION

In recent research on local government efforts have been made to suggest and/or construct policy output or decision-making models that bring together several types of data.[1] This paper seeks to specify such models by broadening the data base so that resource allocation is not simply predicated on previous expenditure patterns or simple relationships based on the cost of a unit of service. According to James Coleman, the need for multiple sources of data is among the principles that should govern policy research.

> For policy research, the criteria of parsimony and elegance that apply in discipline research are not important; the correctness of the predictions or results is important, and redundancy is valuable.[2]

The methodological factors supporting that propostion dictate the use of multiple sources of data and multiple methods of data analysis.[3]

The object of this paper will be to address Coleman's policy principle from an intermediate and necessarily narrower perspective: we review the utility of incorporating survey data into models studying the outputs of local government, stressing questions related to measurement.

The Research support for this paper came from an NSF-RANN grant (SSH-73-030365) to Charles Ries, UCLA Institute of Government and Public Affairs. I would like to thank Sidney Sonenblum, John Kirlin, and Terry Clark for critical comments on an earlier draft of this paper. I am grateful to David Legge for guidance and stimulating comments throughout this research.

EXAMPLES OF POLICY OUTPUT MODELS

Researchers at the Urban Institute have been developing municipal service models conceptualizing productivity measurement in terms of several types of input measurements. Hatry argues that 'for a government to obtain a reasonable perspective of its productivity for any service, it will need mutiple measurements'.[4] He is using the idea of multiple measurement in terms of obtaining several different sorts of measures to build a more complex index of measurement. Hatry's efforts are animated by the fact that he wishes to move beyond simple, single (and most readily available) physical measures.

One of the most ambitious efforts using multiple indicators in measuring the output of local government agencies is that being undertaken by Elinor Ostrom.[5] Ostrom's research goes beyond that at the Urban Institute on two critical points. First is the extent of coverage of municipal services: more than one is examined, which allows for comparative analysis between service domains. Such an analysis allows judgments, for example, about the quality of the data per se as measures of output.

The second point is that Ostrom and her colleagues are undertaking extensive analysis of relationships among indicators within service areas.[6] For example, they will

> . . . ascertain what kinds of relationships exist between the physical measures of service output and citizen perceptions and evaluations of that output. In regard to street lighting, for example, we will examine the relationship between light-meter readings and citizens' perceptions of street lighting on their block. In a relative sense do citizens perceive differences in output as measured with a light meter?

Another project is the effort to formulate a model of community decision-making proposed by Terry Clark.[7] He argues that policy outputs are a function of five variables: citizen values, community characteristics, resources, community leadership, community power and decision-making structures. Clark's model specifies the use of aggregate data (census, voting records, agency reporting information, etc.), structural data (jurisdictional, legal-political), and survey data (elite and citizen).

Clark's research has led him to include citizen attitudes that relate to specific categories of policy outputs.[8] His focus has been on two interrelated areas:

(1) the extent to which citizens desire to increase or decrease the

overall level of local government activities, and

(2) the extent to which citizens prefer to increase or decrease particular local government services.

As in the Ostrom case, multiple measurements of preferences would provide more insight into the psychological meaning as well as the substantive meaning of choice.

THE PRESENT STUDY

This paper is part of a study concerned with the impacts different modes of municipal service provision have on citizen evaluations. The Los Angeles area is an especially good site for such research because several modes of service provision are utilized here. Of special concern to this project are possible impacts of independent cities' contracting for provision of services with the government of Los Angeles County — a type of arrangement developed principally in 1954 and called the Lakewood Plan. Contracting for services is not limited, however, to legal relations between governmental units; cities can obtain services from private firms as well. Like the research undertaken by Hatry, Ostrom and Clark, this project argues that citizen perceptions, preferences, and evaluations can be viewed as an important measure of governmental output or performance, which must supplement other indicators (e.g. economic and physical) utilized as performance measures.[9] Thus, a companion paper,[10] which deals with the substantive aspect of citizen evaluations, specifies regression models employing aggregate and structural data.

MEASURING PERCEPTIONS, PREFERENCES, AND EVALUATIONS

With Webb et al. and Coleman arguing the need for obtaining multiple measures of indicators, there is a necessary convergence of 'discipline' and 'policy' research. Thus, from discipline research, psychometric theory directs the researchers to use multiple measures. A single measure of any concept will contain a number of types of errors — i.e. design, operationalism, a random measurement (error) component. The question that psychometric theory addresses is, to what extent the measured value closely reflects the true value or to what extent it contains substantial error.[11] As Webb[12] and Blalock[13] have argued on the one hand, reliance on a single measure decreases reliability, while on the other hand, it creates data analysis problems. Random error attenuates bivariate correlation and regression coefficients and also produces attenuating biases in multivariate estimates where there is random error in the independent variables.[14]

In an earlier project paper[15] we demonstrated that when a number of indicators are employed (each of which has a random error component, to be sure) and combined into a single index, a more reliable and therefore stronger relationship exists.

From the perspective of policy research, Coleman[16] argues for the use of research designs and methods that give good results with high probability rather than the use of more sophisticated techniques that give excellent results if they are correct. Thus, following Coleman, a model such as employed by Stipak[17] that is severely constrained by sample design and data base problems is nonetheless robust under conditions of only partially met assumptions. The prior conceptualization of the project allows Stipak to use multiple sources of data, while the sample design problem is partially met by using two regression models. Such injunctions again focus on the intimate relationship between considerations of sound methodological decisions and substantive findings.

It is not unreasonable to establish a schema that very simply relates data collection methods and measurement techniques. Such a procedure allows us to classify types of research as well as define what we mean by methodology, methods, techniques, etc. By *methodology,* we mean an elaboration of an entire research strategy: design, sampling, measurement, etc. A *method* is a general rationale employed to specify and structure empirical observations. For example, we would speak of a *data collection method.* Following Leege and Francis,[18] a *technique* 'is a specific instrument for the generation and analysis of data'. Thus, for example, a self-anchoring scale is a measurement technique, while key cluster analysis is a data analysis technique. Figure 1 sets out the schema by which we can classify research.

Perhaps the most prevalent type of research design stipulates only that a single measure or indicator be employed within a single method of data collection. The prevalent use of the single-item indicator demonstrates the problematic nature of the relationship between theory construction and methodology. Newhouse[19] in her literature review points to examples of lack of attention to this relationship in local government research. Leege and Francis[20] make a second argument: that even when validation problems are recognized the use of single indicators as outside criteria in the criterion-validation process is undesirable for dealing with *systematic error.* They further argue, 'Not only the measure of the concept itself, but also the criterion for validation should be based on multiple indicators.'[21]

Figure 1. Research Classification Schema Measurement Technique

Method:	For Single Measure	For Multiple Measures	
		Homogeneous	Heterogeneous
Data Collection			
Questionnaire			
Interview			
Experiment			
Observation			
Physical			
Documents			

An improvement on the single-indicator/single-method perspective is the single-indicator/multiple-method strategy. Here we have a single concept, e.g. quality of solid waste collection, measured by different sources of data. Hatry's research on productivity models would be an example of attempts to obtain two measures of 'quality'. For practitioners' modeling or decision-making needs this is an appealing strategy because models are kept simple and multiple data sources can be more manageable. Such a strategy, however, can potentially suffer most of the problems articulated by Leege and Francis.[22] In effect the strategy is aggregating a set of single-indicator/single-method procedures.

A third strategy is the use of homogeneous mutliple measures with a single data collection method. Leege and Francis[23] call this 'multiple indicators of the same kind', which have played a major role in attitude and personality measurement. They key to this strategy is the use of homogeneous indicators[24] in combination. For example, an underlying trait is hypothesized and a battery of items utilizing the *same* stimulus is administered. All indicators − of the same kind − are then combined to create a new scale score.

The problem is that despite the fact that the 'test' may be relatively homogeneous in toto, the individual indicators may have been subject to biases in either the measurement technique or the data collection method.

A fourth strategy is to employ multiple but homogeneous indicators with more than one data collection method. This will provide an external control for distinguishing error that is peculiar to a data collection method. This is a variation of the Leege and Francis[25] strategy of 'multiple indicators each based on different operations'. In this case we have defined operation as a data-gathering method. Their example fits our need, however.

> . . . if we argue that 'people with a high rate of self-doubting run for political office,' we might administer to a sample of candidates and noncandidates a series of tests where multiple items were designed to measure self-confidence, paranoid tendencies, and need inviolacy. But we might also ask for clinical observations on the sample, or we might design a set of experimental games to measure the same traits. Results of the latter measurement methods (e.g. data collection methods) would help us to estimate to what extent scores on the initial test were a function of the measurement method (sic) employed.

A fifth strategy is to use multiple heterogeneous measurement techniques within a single data collection method. The survey portion of this project represents such an effort. The review[26] undertaken by this project, of citizen orientations in the literature on local government research employing survey methods, demonstrates the lack of rigorous methodological investigation of citizen perceptions, evaluations, and preferences of the outputs of urban governance. Elsewhere[27] we noted that Milbrath, Ostrom and their associates were notable exceptions. Because of a disparate lierture in this area, this project has made preliminary efforts in employing several measurement techniques focusing on policy areas (e.g. city services) in an omnibus survey — the Los Angeles Metropolitan Area Survey (LAMAS).

This strategy is, of course, subject to method effect, as are all surveys. If measurement techniques and their items are operationalized well, then problems of method effect can be tackled. When measures are poorly operationalized and/or the execution of a measurement technique is not thoroughly considered, then we obviously obtain no reading on method effect.

The sixth, and most complete strategy, is the use of multiple heterogeneous measurement techniques and mutiple data collection methods. Such a strategy has been termed *multiple operationalism*. Leege and Francis argue[28]:

> The strategy we espouse for conceptual-operation coordination, then, is one where the concept is carefully explicated and differentiated to derive its location in a theoretical network of concepts. Measures of the concept in all its aspects are developed, as well as measures of other concepts in the

theoretical network; the measures are refined through empirical operations using multiple indicators of the *same* kind and multiple operations of *different* kinds. Then, the entire predictive network surrounding the concept is subjected to empirical examination. Through such procedures . . . theory becomes cumulative and measurement error is less of a mystery.

As suggested by the matrix in Figure 1, we have further differentiated the idea of multiple indicators into homogeneous (e.g. for Leege and Francis, 'the same kind') and heterogeneous types. The latter are characterized by different measurement techniques (e.g. self-anchoring scale, and budget pies.) We then reserve the term 'operations' for various types of data collection methods. For example, certain portions of Elinor Ostrom's current research approximates the most fully developed conceptual-operation coordination.

ORDINAL MEASUREMENT TECHNIQUES
In a previous paper[29] we noted that Robinson and his associates[30] reported very little methodological variation in measurement techniques in three subject domains — domestic government, community-based attitudes, and political participation. Of the nineteen scales reported, only ordinal or dichotomous (agree-disagree) measures had been employed, with very little accompanying data on validity and reliability.

Ordinal scale measures appear to be the most popular because they are easy to administer, whether by interview or self-administered questionnaire, and simple to score. In policy research such techniques continue to be employed. Therefore, we used two ordinal techniques for rating specific local government service areas, and a third only for police service.

Certain parallels between the present project and that of Everett Cataldo et al. in Buffalo deserve mention.[31] Cataldo and his associates reported that the card-sorting technique proved superior from both 'quality' and 'operational' perspectives. From the LAMAS field effort, the operational points (ease and time of administration, respondent enjoyment, etc.) support the Buffalo findings. We differ in terms of the 'quality' considerations. This outcome seems due to the fact that the number of measures used in Buffalo was *substantially* greater than those in the LAMAS survey. Secondly, the Buffalo 'out-take' scale was composed of many more heterogeneous items than the service items on LAMAS.

Four pieces of information for the LAMAS VII data are relevant

Table 1. Citizen Evaluations and Preferences

	Card Sort						Ordinal						Budget Pie					Self-Anchoring Scale
	Street Repair	Police Services	Parks & Recreation	Street Cleaning	Garbage Collection	Construction Control	Street Repair	Police Service	Parks & Recreation	Street Cleaning	Garbage Collection	Construction Control	Street Repair	Police Services	Parks & Recreation	Street Cleaning	Garbage Collection	Police Services
Card Sort																		
Street repair																		
Police services	0.15																	
Parks & recreation	0.17	0.25																
Street cleaning	0.26	0.12	0.14															
Garbage collection	0.21	0.16	0.12	0.33														
Construction control	0.18	0.15	0.31	0.13	0.12													
Ordinal																		
Street repair	0.62	0.10	0.12	0.25	0.15	0.17												
Police services	0.19	0.62	0.19	0.13	0.13	0.18	0.23											
Parks & recreation	0.19	0.22	0.47	0.11	0.10	0.30	0.23	0.27										
Street cleaning	0.26	0.13	0.10	0.23	0.27	0.10	0.38	0.25	0.24									
Garbage collection	0.18	0.13	0.10	0.23	0.62	0.10	0.26	0.22	0.19	0.44								
Construction control	0.11	0.14	0.25	0.06	0.09	0.69	0.25	0.21	0.33	0.11	0.18							
Budget Pie																		
Street repair	0.09	0.09	0.02	0.02	-0.02	0.01	0.11	0.07	0.00	-0.03	-0.03	0.04						
Police services	0.02	0.01	0.05	0.05	0.01	0.03	0.11	0.06	0.04	0.03	0.03	0.05	0.31					
Parks & recreation	0.00	0.04	0.22	0.02	0.06	-0.12	-0.02	-0.02	0.19	0.01	0.00	0.10	0.20	0.17				
Street cleaning	0.01	0.07	0.03	0.03	0.01	0.02	0.01	-0.06	-0.02	-0.05	0.00	0.08	0.54	0.37	0.21			
Garbage collection	0.00	0.09	0.04	0.04	0.03	0.01	0.01	0.05	-0.01	-0.02	0.00	0.06	0.54	0.35	0.19	0.70		
Self-Anchoring Scale																		
Police services	0.14	0.55	0.21	0.16	0.11	0.11	0.16	0.64	0.27	0.18	0.17	0.13	-0.03	0.06	-0.04	-0.05	-0.05	

to its quality. Two types of data analysis were employed, cluster analysis[32] and a multi-trait/multi-method matrix (Table 1). The underlying assumption of such a matrix is that, when two techniques are compared, correlations of the same measures will be substantially larger than correlations among different measures. For example, the correlation between the street repair items on both techniques (e.g. card sort and ordinal) is 0.62, whereas the correlation between street repair and other service measures *within and betweeen techniques* is never more than 0.26 (e.g. its correlation with street cleaning). Thus, in interpreting such a matrix, a validity diagonal is established by the correlations obtained on similar measures between the two techniques. Where a panel study has been used or items have been apportioned on an instrument in a split-half manner, then a reliability diagonal can be designated by correlations of similar measures *within the same technique.*

Given that neither condition was possible with the LAMAS survey, a second data analytic strategy — the use of cluster analysis[33] — was adopted. A cluster analytic solution on its face appears to be analogous to factor analytic solutions because both consist of a number of dimensions or clusters defined by subsets of observed variables. The desired objective is to make each dimension as generally mutually exclusive of the others as possible. Both types of solutions are specified by loadings of an observed variable on dimensions as well as intercorrelations among the dimensions. It is at this point that the two types of solutions employ dissimilar methodological criteria for establishing dimensions. In factor analysis it is sufficient for an observed variable to be highly correlated with other variables in the dimension. A cluster analytic solution imposes a more exacting criterion before admitting a variable to a cluster or dimension: it must have a similar pattern of relationships with all the variables in the *predefined* cluster. Thus, accompanying a matrix of loadings of the observed variables on the cluster is a second matrix of similarities that exhibits the extent to which a variable has a pattern of correlations that is similar to those of the items which are preselected to define the cluster. The similarity coefficient ranges from zero to one, indicating a total lack of similarity (0.0) to a perfect fit with the cluster items (1.0).

A third coefficient, Cronbach's Alpha, is employed because it measures the reliability of the scale formed by its defining variables. A reliability coefficient indicates how much of the total variance of a measured variable is 'true' variance. Thus, a high reliability coefficient means that the measure is relatively free of error variance and contains only uncorrelated measurement error.

Thus, by using a cluster analytic solution with the loading, similarity, and reliability coefficients one can link substantive findings to methodological considerations (e.g. validity, reliability, and technique comparison). Table 2 reports the above coefficients for scales created from 17 service evaluation items. Eleven scales were constructed, of which six were defined only by card sort and ordinal items. In each of the six scales[34] both technique items had identical loading and matrix coefficients. In no case was there another item in the scale that exhibited a higher value. In all but one scale, police services (no. 6), the loading matrix entries were substantially greater than all other entries, thus quantitatively defining the dimension. When coupled with identical similarity matrix entries within service scales, a strong content validity case is established for individual service scales.

In the case of the police services scale (no. 6), the card sort and ordinal items exhibit the same pattern as in the other scales, but the inclusion of a self-anchoring scale item (no. 9, Q55A) rating police services shows quite a similar pattern on scale 7. Therefore, when we specifically define a scale (no. 8) with all three measurement technique items, we note that both loading and similarity matrix entries are nearly identical for all three items and that they indeed define a better (i.e. a more error-free) police services cluster.

The relevance of a cluster analytic solution for validity and reliability concerns is amply demonstrated by the differences in the police service scales 6 and 8. The card sort/ordinal scale has a reliability coefficient of 0.77, but when the self-anchoring scale item was introduced, an 0.82 reliability coefficient was obtained. However, when two additional police service evaluation questions using a comparative judgment technique are included in scale 7, the reliability coefficient is only 0.77. The spread of the similarity coefficients is from 0.78 to 0.90, not great, but the corresponding range for loading matrix coefficients is 0.41 to 0.81. This suggests that the content validity of the items is not as strong as either scale 6 or scale 8.

If we define a scale (no. 9) by the two comparative judgment questions, we note identical loading and similarity coefficients, but the other police service items exhibit the equivalent characteristics found in scale 7 (i.e. all police items). That the comparison items do not form a separate cluster is suggested by the loading and similarity coefficients, but *most specifically* the very low reliability coefficient

Table 2. Citizen Evaluation of (and Preferences for) Municipal Services Loading and Similarity Matrics

2 Street Repair	3 Street Cleaning	11 Street Repair & Cleaning	1 Streets All Items	4 Parks & Recreation	5 Garbage Collection	6 Police I	8 Police II	9 Police Comparison	7 Police All Items	10 Construction Control	Item No.		
79 90	33 69	60 76	59 79	22 46	25 55	22 42	21 42	15 31	20 38	18 43	1	Q25A	Rating of Street Repair
22 15	17 69	25 72	29 74	13 54	21 66	20 52	19 52	04 46	16 50	13 48	2	Q25B	Rating of Traffic Signals
22 41	14 30	23 35	23 39	37 63	14 28	21 35	19 36	16 28	19 33	83 90	3	Q25C	Rating of Construction Control
16 39	16 38	20 39	20 41	29 51	18 32	28 49	76 23	52 82	67 87	18 35	4	Q25D	Rating of Police Services
19 43	17 35	22 39	23 41	82 93	14 29	19 40	28 49	23 36	28 44	34 61	5	Q25E	Rating of Parks & Recreation
33 66	79 90	60 74	56 74	15 38	37 69	18 33	20 39	16 30	20 35	11 29	6	Q25F	Rating of Street Cleaning
37 68	37 68	38 60	38 60	14 28	79 87	37 77	17 34	13 23	17 30	13 25	7	Q25G	Rating of Garbage Collection
05 22	07 25	08 23	08 25	02 27	09 22	75 96	37 78	50 77	44 78	04 17	8	Q51	Police: Own vs. Other Neighborhoods
19 41	21 40	25 40	27 42	30 49	18 37	45 88	77 96	64 34	72 90	14 36	9	Q55A	Rating Police Services Today
13 40	13 37	16 39	19 40	18 55	08 30	18 53	46 88	50 77	52 85	16 41	10	Q57	Police: Own vs. Adjacent City
38 87	27 74	40 80	43 80	22 46	23 63	21 41	19 51	17 38	19 46	19 49	11	Q58A	Rating of Street Lighting
29 90	40 67	66 75	66 77	25 40	26 57	22 35	21 41	14 31	20 37	25 41	12	Q58B	Rating of Street Repair
23 43	11 33	21 38	25 40	35 62	17 28	20 33	20 35	17 30	20 33	83 90	13	Q58C	Rating of Construction Control
27 43	27 42	34 43	36 45	28 51	22 38	79 93	81 95	80 83	81 89	23 36	14	Q58D	Rating of Police Services
28 59	44 73	45 66	46 66	18 34	79 87	22 37	22 37	13 29	21 34	17 30	15	Q58E	Rating of Garbage Collection
41 70	79 90	67 76	65 76	22 39	45 72	24 41	24 41	16 32	23 38	13 34	16	Q58F	Rating of Street Cleaning
27 49	22 42	30 45	32 48	82 93	20 34	31 53	33 51	21 43	31 48	38 64	17	Q58G	Rating of Parks & Recreation
-02 06	-03 03	-01 04	-01 05	01 03	04 02	-10 04	-04 06	12 03	01 05	-14 09	18	Q24A	Budget Pie Allocation Parks & Recreation
00 03	-03 02	-10 03	-10 03	01 03	-03 01	-09 04	-09 03	03 01	-05 02	03 00	19	Q24B	Budget Pie Allocation Street Repair
-12 02	-09 01	-02 02	-02 02	-03 02	-02 00	-08 04	-07 03	-01 01	-06 02	03 00	20	Q24C	Budget Pie Allocation Garbage Collection
01 02	-05 01	-05 01	-05 01	-04 02	-00 00	04 00	-08 03	-04 00	-07 02	06 00	21	Q24D	Budget Pie Allocation Street Cleaning
00 01	05 01	03 01	04 00	06 00	02 02	06 03	05 01	12 03	08 02	05 02	22	Q24E	Budget Pie Allocation Police Services
06 03	02 02	05 02	06 02	12 06	-03 01	04 03	04 03	-02 01	02 02	-01 02	23	Q24F	Budget Pie Allocation Own Tax Reduction
0.77	0.77	0.73	0.72	0.81	0.77	0.77	0.82	0.39	0.77	0.82		Reliability Coefficients	

Key: 1st entry = Loading Matrix; 2nd entry = Similarity Matrix; Q.25A–Q.25G = Card Sort; Q.58A–Q.58G = Ordinal.

Table 3. Citizen Evaluations of Municipal Services: Scale Intercorrelation Matrix of Ordinal Measurement Techniques

	2 Street Repair	3 Street Cleaning	11 Street Repair & Cleaning	1 Streets All Items	4 Parks & Recreation	5 Garbage Collection	6 Police I	8 Police II	9 Police Comparison	7 Police All Items	10 Construction Control
Street repair (1, 12)* 2**		0.36	0.82	0.81	0.22	0.25	0.21	0.21	0.10	0.19	0.21
Street cleaning (6, 16) 3			0.82	0.78	0.18	0.40	0.21	0.22	0.11	0.21	0.12
Street cleaning & repair (1, 6, 12, 16) 11				0.96	0.24	0.39	0.27	0.26	0.13	0.24	0.20
Streets all items (1, 6, 11, 12, 16) 1					0.26	0.40	0.27	0.28	0.14	0.26	0.22
Parks & Recreation (5, 17) 4						0.16	0.28	0.30	0.15	0.28	0.35
Garbage collection (7, 15) 5							0.20	0.20	0.09	0.18	0.15
Police I (4, 14) 6								0.96	0.46	0.88	0.15
Police II (4, 9, 14) 8									0.48	0.92	0.19
Police comparison (8, 10) 9										0.78	0.11
Police all items (4, 8, 9, 10, 14) 7											0.19
Construction control (3, 13) 10											

*   Numbers in parentheses are item numbers;
**   Numbers 1-11 outside parentheses are scale numbers.

of 0.39.

It is also interesting to examine relationships between scales. Table 3 presents the scale intercorrelation matrix which demonstrates how mutually exclusive one scale is from the others. For example, it can be seen that if the police comparison scale (no. 9) is correlated (0.46 and 0.48) with the police ordinal measurement scales (nos. 6 and 8), they are obviously tapping different aspects of a law enforcement domain.

On balance then, we note that the card sort and ordinal methods both measure service-specific areas and provide valid data. There is, however, one aspect in each analysis method – cluster analysis and multi-trait/multi-method matrix – that upon further investigation may more fully support Cataldo's finding that the card-sort technique provides more reliable and valid data. In multi-trait/multi-method matrix, when comparing the within-method intercorrelations (e.g. the heterotrait or heteroitem-monomethod submatrices), all coefficients within the ordinal technique are larger than in the card-sort technique except one (Construction Control). Such correlations within techniques should be low, otherwise one might suspect method effects (e.g. yea-nay saying and position effect).

An equivalent perspective is an inspection of Table 2 to examine the similarity coefficients within and then across the two techniques. Each technique across all 11 scales includes 66 similarity coefficients. When compared, 54 coefficients of the ordinal technique are larger than their card-sort equivalents. The larger the similarity coefficient, the closer it comes to the largest scale-defining items. Thus, like the multi-trait/multi-method matrix, there tends to be a higher, more positive set of inter-item correlations in the ordinal technique than in the card-sort technique. Because there are only six equivalent items in each technique, it would be hasty to conclude that a method effect has been uncovered. If a larger number of items had been included, it may well have led to response bias in ordinal scales.

THE RELATIONSHIP BETWEEN EVALUATIONS AND PREFERENCES: THE BUDGET PIE

Several authors have recently argued that a budget-pie measurement technique should be used. The central area of agreement about its utility is that respondents should be forced to 'trade-off' some of their preferences against others.[35] Traditional ordinal measurement techniques are criticized because they can lead to reporting preferences for more of each service. By contrast, the utility of the budget pie, Clark argues,[36] is such that intensity of preference is expressed under conditions of budget constraint — a factor only too obviously operating in the real world of local government decision-making. Hoinville and Clark both suggest that preferences must be examined in clusters so that citizens' responses are in the form of tradeoffs.[37]

The idea of obtaining opinions and preferences within a constrained decision-making context to date has been employed infrequently.

The LAMAS respondents were given a blank budget pie equivalent to one that showed current service allocations and were then asked to draw in new lines, given their preferences *within the current budget.* Table 4 shows the distribution of Los Angeles citizens 'reallocating the budget pie'.

Table 4. Changing the City Budget (Sample Size 1028)

Changers	66.6%
Status Quo	30.2
Don't know	1.9
Refused	1.3
	100.0%

At the outset of the survey there was concern about whether people would have the cognitive skills and motivation to be able to 'cut the budget pie'. The distribution suggests that very few people either flatly refused or admitted not knowing. However, 30.2 percent of the sample retained the current allocation (i.e. status quo) as their preference. Beardsley's research was a two-wave panel study, and he did not report status quo figures for either of the cross sections. However, one can extrapolate from his findings based on both waves and locate at least 9.9 percent of his respondents maintaining a status quo preference.

The sample design prohibits extended analysis of the impact of jurisdiction; but we may ask if residents of particular categories of cities or areas within Los Angeles City were 'budget changers'. Table 5 maintains some of the regional divisions embedded in the LAMAS sample while aggregating into a separate category citizens in various cities that engage in some form of contracting. We note that residents of Long Beach, contract cities, and northeastern cities reported the

Table 5. Budget Status by Area of Los Angeles City or Type of Jurisdiction

| Budget Reallocation | Independent Cities | | | | Los Angeles City | | | | | |
	N.E. Cities	Western Cities	S.&E. Cities	Long Beach	Contract Cities	West L.A.	S.F. Valley	South L.A.	Central L.A.	L.A. County Unincorp
Changers	52.1	68.1	67.6	43.1	50.8	71.9	75.2	90.1	95.5	72.4
Status Quo	47.9	31.7	32.4	56.9	49.2	28.1	24.8	9.9	6.6	27.6
	100.%	100.%	100.%	100.%	100.%	100.%	100.%	100.%	100.%	100.%
N	96	82	108	51	132	121	129	71	89	116

highest proportion of status quo responses.

The Los Angeles respondents were given a second blank budget pie and told that each household in their city could allocate $100 of new money to the five service areas or could take all or some of that money in a tax reduction. 3.9 percent either refused or had no opinion on this item.

How do allocation budget pie results compare to those with the earlier formats? In terms of the multi-trait method matrix and cluster solutions of the previous section, it seems that service evaluations using the card sort, ordinal scale, and self-anchoring scale are not related to preferences on the $100 allocation budget pie.

In the multi-trait/multi-method matrix (Table 1), the validity diagonals show correlations near zero on all three evaluation methods. Only the parks and recreation measure exhibits a very modest (negative) relationship (-0.22, card sort; -0.19 ordinal scale).

What the multi-trait/multi-method matrix demonstrates with this limited number of services is the nature of the budget pie as a preference measurement technique. Citizens tend to prefer to allocate resources in terms of broader areas of service. The correlation between street-cleaning preference and garbage collection is 0.70. Street repair is related to both street cleaning and garbage collection at the 0.54 level. The inter-item evaluation correlations exhibit no equivalent strength among broader service domains.

NOTES

---

1. H. P. Hatry and D. M. Fisk, Improving Productivity and Productivity Measurement in Local Governments (Washington, D.C.: The National Commission on Productivity 1971); H. P. Hatry, 'Issues in Productivity Measurement for Local Government', Public Administration Review, Vol. 32 (November/December 1972); T. N. Clark, Community Power and Policy Outputs (Beverly Hills and London: Sage 1973); E. Ostrom, 'The Need for Multiple Indicators in Measuring the Output of Public Agencies', Policy Studies Journal,

Vol. 2 (2), 87-92; J. E. Grigsby III et al. Prototype State-of-the-Region Report for Los Angeles County (Los Angeles: School of Architecture and Urban Planning, UCLA); V. Green et al., 'The Measures Project – A Theoretical and Methodological Overview', Research Report Number 1, Workshop in Political Theory and Policy Analysis, Indiana University, n.d.

2. J. S. Coleman, Policy Research in the Social Sciences (Morristown, N.J.: General Learning Press 1972), 4.

3. Coleman, op. cit., 5.

4. Hatry, op. cit., 783.

5. Ostrom, op. cit.

6. Ibid.

7. Clark, op. cit.

8. T. N. Clark, 'Please Cut the Budget Pie', paper 37 of the Comparative Study of Community Decision-Making, University of Chicago; T. N. Clark, 'Can You Cut a Budget Pie?' Policy and Politics, Vol. 3 (December 1974), 3-31.

9. B. Stipak, Citizen Evaluations of Municipal Services in Los Angeles County (Institute of Government and Public Affairs, UCLA 1974).

10. Ibid.

11. H. Asher, 'Some Problems in the Use of Multiple Indicators'. A paper presented at the conference on 'Design and Measurement Standards for Research in Political Science', Lake Lawn Lodge, Delevan, Wisconsin, 13-15 May 1974.

12. E. J. Webb, Unobtrusive Measures: Non-Reactive Research in The Social Sciences (Chicago: Rand McNally 1966).

13. H. M. Blalock, Jr, 'The Measurement Problem: A Gap Between The Language of Theory and Research', in Hubert M. Blalock, Jr and Ann Blalock (eds), Methodology in Social Research (New York: McGraw-Hill 1968), 5-27.

14. Asher, op. cit.; Blalock, op. cit.

15. D. Scott, 'Contracting Project Survey: Some Methodological Considerations' (Institute of Government and Public Affairs, UCLA Mimeo 1974).

16. Coleman, op. cit., 5.

17. Stipak, op. cit.

18. D. D. Leege and W. L. Francis, Political Research (New York: Basic Books 1973).

19. M. Newhouse, 'Literature Review and Propositional Inventory of Citizen Orientations and Evaluations' (Institute of Government and Public Affairs, UCLA mimeo 1974a).

20. Leege and Francis, op. cit., 139.

21. Ibid.

22. Ibid.

23. Ibid.

24. Asher, op. cit., 2.

25. Leege and Francis, op. cit., 140.

26. Newhouse, op. cit.

27. Scott, op. cit.

28. Leege and Francis, op. cit., 141.

29. Scott op. cit.

30. J. P. Robinson et al., Measures of Political Attitudes (Ann Arbor: Institute for Social Research, University of Michigan 1968).

31. E. F. Cataldo et al. 'Card Sorting as a Technique for Survey Interviewing' Public Opinion Quarterly, Vol. 34 (Summer 1970), 202-15.

32. C. Hensler, 'The Structure of Orientations Toward Government' (Unpublished Ph.D. dissertation, MIT. Cambridge, Mass. 1971); R. C. Tryon and D. E. Bailey, Cluster Analysis (New York: McGraw-Hill 1970).

33. The following discussion draws heavily on Hensler op. cit. 69ff.

34. Scales numbered 2, 3, 4, 5, 6, and 10.

35. G. Hoinville and R. Berthoud, 'Identifying Preference Values: A Report on Development Work', (London: Social and Community Planning Research 1970); P. L. Beardsley et al, Measuring Public Opinion on Priorities (Beverly Hills and London: Sage Professional Papers in American Politics 04-014, 1974). Clark, 'Can You Cut . . .'; Scott, op. cit.

36. Clark, 'Please Cut . . .'.

37. P. Cadei, 'Equity and Responsivenesss in Local Government Performance' (Institute of Government and Public Affairs, UCLA, mimeo. 1974).

Citizen Surveys for Local Governments: A Copout, Manipulative Tool, or a Policy Guidance and Analysis Aid?

Harry P. Hatry and Louis H. Blair

More and more local governments in the United States appear to be sponsoring surveys of randomly selected samples of the community population. For the most part the surveys have been on a one-time basis for some special purpose (such as for transportation studies or for housing condition inventorying). Only rarely have such surveys been undertaken at periodic intervals with similar questions so that the effects of changes in policies and programs can be assessed and that time trends be identified. The research community, the federal government, national polling firms and some universities have ongoing, periodically repeated surveys such as the current population survey and health interview survey, which do obtain time series information on a variety of topics. Properly administered surveys conducted annually are a potentially very useful management tool for local governments.

USES FOR CITIZEN SURVEYS
Why should a city or county government undertake surveys, particularly on a regular basis?[1] The citizen survey has a very appealing characteristic. It is a means for obtaining citizen inputs into government programs and policies that is probably considerably more representative of the whole population than inputs from other sources (except, perhaps, elections) and without the volatility of information sometimes present in other approaches. Other forms of citizen participation such as community meetings, hearings of city or county councils, daily meetings by local officials with members of the public, and volunteered citizen complaints, all have significant problems of potential non-representativeness and sometimes involve direct confrontation with angry citizens. Generally citizens who participate are persons who are most vocal, who understand the government system, and who have the

time and energy to make their feelings known. These do not seem likely to be the typical citizen. The citizen survey on the other hand, even with small samples of the population (if the sample is drawn properly) obtains opinions from persons more representative of at least the major part of the population (though, even the most elaborate citizen surveys inevitably will miss certain hard-to-reach portions of the population).

Two major categories of survey uses need to be discussed.

(1) The first and the most common use of citizen surveys by local governments has been to obtain opinions from citizens as to their preferences and priorities relative to specific services or policies. A typical question of this type is 'How would you rank the following city services in terms of their satisfactoriness?' A similar usage is to ask the citizens for their opinions on certain specific forthcoming policies such as 'Do you believe the laws on victimless crimes such as gambling and sexual acts between consenting adults are: over-enforced; under-enforced; or enforced about right?' In essense, this type of question is generally used to directly help a government set priorities, develop plans or determine policies. The Dayton (Ohio) Public Opinion Center working with the Dayton City Government was a leader in the early 1970s in conducting surveys of this type. The Columbus (Ohio) City Council has used such a survey to obtain citizen inputs on the expenditure of revenue sharing funds. The City of Richmond, Virginia, in conjunction with the Bureau of Social Science Research conducted an extensive survey of this type.[2] Many other communities have on occasion also undertaken such surveys.

(2) The second general category of usage is to provide background and analytical data to a government for use in planning and for establishing or revising programs and policies. This usage has appeared to be considerably less in use in the United States. We suspect that this usage has the greatest, currently untapped benefit. This category includes a variety of types of questions such as:

(a) Questions which provide 'factual' information (e.g. estimates of the frequency of unreported crimes; reasons for non-reporting; frequency of rat sightings in the neighborhood; frequency of missed garbage collection; and estimates of the number and percentage of *different* households that have used such services at libraries; recreation facilities or public transit). Such information can be obtained from citizen surveys, often more easily than by other means.

(b) Questions to provide indicators of the effectiveness or quality of government services based on citizen perceptions of the satisfactoriness of various aspects of those services (e.g. citizen perceptions of the cleanliness of their streets, their feeling of security, the adequacy of street lighting in their neighborhoods, the adequacy of recreation and library facilities and programs, the helpfulness and courteousness of local government employees, and timeliness of the responses by government employees to their request for services or information). Those who feel that citizen perceptions as to the satisfactoriness of their services, are equally, or even more relevant for measuring government performance than more 'factual' evidence (such as how clean streets 'really' are, reported crime rates, etc.) will be particularly attracted to the citizen survey for obtaining such information.

(c) Questions on why citizens do not use public services or facilities, separate answers into conditions over which the local government has some control (hours of operation, accessibility, cleanliness, helpfulness of staff) and conditions over which it has little or no control (citizens are not interested, use private facilities, are too old, etc.).

Examples of questions that might be included in a citizen survey for this category of usage are shown in Figure 1.

Related information from public records, such as crime rates and numbers of citizen complaints are also relevant for assessing services, but such information has problems more severe than is commonly recognized. For example, reported crime rates and citizen complaints tend to underestimate dissatisfaction — many citizens won't call the government, don't know how to, or are afraid to, or believe it wouldn't do any good if they did. Information on service performance from the various sources should be considered as complementary to each other, each probably assessing a somewhat different characteristic. Thus, for example, citizen perceptions of the cleanliness of their neighborhoods complement data from trained observer ratings using a pre-selected photographic rating guide of street cleanliness. If the two sources do not appear to be in close agreement, which often appears to be the case, this does not mean that the information from one or the other is invalid — at least for those who believe that citizens' feelings are important regardless of whether conditions are in some physical sense better or worse than rated by the citizens.

The Planning Department for the City of Detroit has had one of the most ambitious local government citizen survey programs to collect demographic data and physical facility data for planning

Figure 1

Typical Survey Questions*

1. In general, would you rate the St Petersburg Bus Service AVAILABLE to you and members of your household as Excellent, Good, Fair or Poor?

2. Would you say the amount of street lighting at night in this neighborhood is about right, too low (need more lighting), or too bright (more lighting than necessary)?

3. Generally speaking, would you rate the St Petersburg Municipal Library Services as Excellent, Good, Fair or Poor?

4. I am going to read a list of reasons SOME people have given for NOT using libraries more often. Please tell me which, if any, generally are TRUE for you or members of this household?

    Library hard to get to
    Library not open right hours
    Personal health problems prevent my using the library
    Buy my own books and magazines
    Not interested in library
    Too busy to go to library
    Poor staff service at library
    Library does not have books I want.

5. Turning now to neighborhood cleanliness, would you say your neighborhood is USUALLY Very Clean, Fairly Clean, Fairly Dirty or Very Dirty?

6. In the past 12 months, did the collectors ever miss picking up your trash and garbage on the scheduled pick-up days? (IF YES, ASK:) How many times would you say this occurred?

7. Turning now to police protection and public safety, how safe would you feeling walking alone in this neighborhood at NIGHT, Very safe, Reasonably Safe or Not Safe at All? (Repeat for 'DAY').

8. In the past 12 months, did anyone break in or was there strong evidence someone tried to break into your home? (IF 'YES', ASK:) How many times did this occur? Were all incidents reported to St Petersburg Police? (IF 'NO' ASK:) How many were not reported? What was the main reason for NOT notifying the police?

---

*Extracted from the 1975 St Petersburg Citizen Survey. Administered by Suncoast Opinion Survey for the City of St Petersburg, Florida during October 1975 purposes. Four percent of the households (about 18,000) were interviewed in 1965; 2 percent in 1969; and another 4 percent will be interview between 1975 and 1977. In recent years a growing number of jurisdiction such as St Petersburg, Florida; Nashville, Tennessee; Dallas, Texas; Palo Alto, California; and Randolph Township, New Jersey have undertaken surveys primarily aimed at obtaining this second category of information — feedback on the quality of their services.[3,4] However, there are important problems and limitations with citizen surveys. These are discussed in the following sections.

CITIZEN SURVEY PROBLEMS — COSTS AND TECHNICAL LIMITATIONS

Cost is probably the key barrier to the more frequent use of citizen surveys by local governments. Survey costs for the type of surveys discussed here are likely to vary from as low as about $5,000 to $10,000 to perhaps $50,000 or more for lengthy complex surveys of a large number of citizens. A primary cost factor will be the number of persons interviewed; it will be determined in large part by the level of precision sought and the number of different clientele groups, districts or neighborhoods of the jurisdiction for which individual sets of statistics are desired. Most governments can obtain sufficient accuracies for many program and policy purposes by interviewing perhaps only 100 households per district — or about 300 to 1500 households as were interviewed in the previously mentioned cities (except for Detroit). There will be potential inaccuracies with such small samples; they may, for example, permit officials to have the confidence that with, say, a 90 percent probability, the results will be within 5-7 percentage points of what a complete enumeration of the population would have found. But such confidence ranges for the resulting data often will be quite acceptable in governments where little information on program performance is otherwise available and that which is available is subject to considerable uncertainty. (This is probably the typical situation.) St Petersburg and Nashville incurred cost outlays of approximately $12 per interview, but in addition government staff spent several person-months preparing the questionnaires and analyzing findings.

Out-of-pocket expenditures can be reduced by the use of volunteer help (for example League of Women Voters personnel such as used in the Randolph Township, New Jersey survey) or existing government personnel for interviewing, if adequate training and supervision is provided for interviewers. Even though this will usually be only a fraction of one percent of a government's total expenditures, such funds may be difficult to obtain. With the magnifying glass of public attention focused on them, such funds are quite vulnerable to being considered non-essential expenditures and dropped by city and county councils and executives during lean periods (which seem to be frequent).

It is beyond the scope of this article to delve into the technical aspects of survey sampling procedures. However, there are probably even more important technical problems than sample size which will affect the accuracy of the findings. For example:

It is often quite hard to locate certain parts of the population, particularly disadvantaged minorities;

persons may not be at home when interviewers call or telephone, and when they are contacted, may refuse to be interviewed;

wording of the questions can be ambiguous and even misleading, thereby unintentionally biasing responses, unless considerable care is taken in formulating the wording;

respondent's memory limitations can affect the accuracy of data such as past victimization or data on past uses of public facilities;

some types of citizens may not be able to respond for themselves such as: non-English speaking residents, the very young, the senile, or those with other types of disabilities.

Sample surveying is by no means a precise science, though it features many sophisticated and technical features. Some of the national polling firms can recite instances of considerable past successes such as the Gallup Polls' past pre-election presidential polls. These organizations also occasionally have their problems such as a major discrepency between the findings of two major pollsters in assessing the population's preferences for specific Republic versus Democratic Presidential possibilities in late 1976 — based on polls taken within one to two weeks of each other. The pollsters principal response was to suggest that the discrepancy was primarily due to a fast turnabout in the publics' viewpoints caused by events during that time period.[5]

Such technical problems suggest that users of survey results should not take them to be exact estimates of the views of the entire population (the non-sampling errors are often likely to be more of a problem than the possible statistical sampling errors which are a function primarily of the sample size). Nevertheless, in government decision-making the information base *generally* available to public officials is so imprecise that it would appear that the information obtainable from citizen surveys, despite the numerous technical problems is likely to be *more* precise than most other sources of information. The classic example is reported crime rates used as a measure of total crime in a community. Despite their seeming precision, these appear to be less accurate than estimates coming from properly done citizen surveys.

There is one final *combined* technical and cost problem. To keep costs down local governments may be tempted to use mail surveys, perhaps sending them along with water bills or as inserts to a local newspaper or newsletter. This gives high exposure and large numbers of returns — much more than normally obtained in a sample survey undertaken by in-person or by telephone interviewing. But the usual low response rates, typically 25% or considerably less, prevent such findings from being credible or being representative of the population. Recent directions in survey work, to reduce costs, have been towards various combinations of the three interview methods, such as to follow-up non-responses to a mailed questionnaire, perhaps by telephone or at-home interviews to achieve respectable response rates.

CITIZEN SURVEY PROBLEMS — COPOUTS AND MANIPULATIONS

The author's opinion is that the following problems are more serious than the foregoing technical problems in the uses of citizen surveys.

Copouts

Public officials sometimes appear too quick to use the citizen survey as a pseudo-referendum to direct or to justify government public actions.

In too many cases citizens are asked questions for which most of them do not have sufficient information and experiences to give informed or meaningful answers. This applies to the spectrum of questions asking the citizen what the government 'ought to do' regarding specific public services. Such a question as 'Do you believe that a government should put more or less resources into the police department?' is in fact a very complex question. To make an informed response requires a knowledge of the costs and benefits likely to result from such actions, information which the average citizen has little likelihood of having. Although information on such questions is often also poor for public officials, they do have access to at least some relevant information and clearly have been assigned the responsibility for making such choices. The only question of this type that the authors have felt to be satisfactory was one asked by a southwest city which asked their citizens whether they would be willing to pay an additional $2.50 per month to obtain backdoor rather than curb collection. In such a case the citizen respondent is, in a sense, an expert; he knows the major cost to him ($2.50 per month) and for the most part can identify the major outcomes of the decision (backdoor versus curbside collection of his garbage).

This problem also applies to the type of question in which the citizen is asked to rank or help set priorities on each of various services. The budget-pie approach discussed elsewhere in this issue is an attempt to force the respondent to consider an overall budget constraint when assessing individual programs or services. However, as with other priority-determining questions, it too has the major problem that the respondent generally has little meaningful knowledge as to the existing tradeoffs between funds applied to one program or service versus another. Thus, in effect, such questions probably only provide information as to the citizen's gut feelings as to the satisfactoriness of the individual services. Responses to such questions should not be considered to provide rational, specific, allocation decisions. Unfortunately, users of the survey data may be tempted to read too much into the responses to such questions.

The Regional Planning Association in the New York City area has attempted to attack this problem by providing extensive information to citizens on selected issues prior to surveying them on these issues.[6] If citizens to be surveyed can be reached in such a manner, permitting them to make informed responses on issues relevant to final choices, then perhaps such questions can be more meaningful. Without this, it would seem that surveys should be limited to less potentially misleading questions that do not tempt the user of the survey results to read too much into the responses.

Because of the potential effects of such questions, another problem sometimes arises. Local government officials are sometimes concerned that citizen surveys might tie their hands. That is, the findings of such surveys would be a source of political pressure, forcing officials into taking some action which other information leads them to feel was inappropriate. This problem seems most apt to occur with questions asking for citizen opinions on what the government ought to do, and thus appear to 'mandate' a particular government action, rather than with questions that inquire about citizens' experiences with the performance of government services.

Manipulations

Survey questions are also subject to manipulation. The choice of working or the inclusion or exclusion of certain options such as those used in rating or ranking questions can lead to almost any desired results. Leading questions are another form of such manipulation.

WILL THE SURVEYS THEMSELVES AFFECT CITIZENS ADVERSERLY OR POSTIVELY?

First, will citizens feel bothered or annoyed at being asked for the interview? The evidence we have to date indicates that citizens are generally willing to spend periods of at least up to about 30 minutes on reasonable questions involving their own personal experiences with local government services. A St Petersburg, Florida survey firm found that citizens appeared highly interested in the government-sponsored survey. If anything, the problem was to complete the interview tactfully without excessive conversation. (To some extent this may be due to the large proportion of elderly in St Petersburg who apparently enjoyed talking to the interviewers.) However, refusal rates and non-completion rates in most of the surveys of the cities identified earlier were quite small. In one case the refusal rate went as high as 12 percent overall. The interviewers felt refusals were due primarily to the fear of letting strangers into their homes. The organization that undertook that at-home survey is considering a combination of mail and at-home surveys and possible telephone surveys in the future.

A second issue is the increasing concern among national polling firms that the population of the United States is becoming over-sampled by the combination of organizations sponsoring citizen surveys to a point where citizens are beginning to resist interviews. This problem has not appeared to have surfaced yet in the government surveys on public services that we have observed. The small samples used by most cities (less than 1 percent of the households were interviewed in St Petersburg, Nashville and Dallas) imply that no household is likely to be interviewed by the government more than once every generation even if the local government undertakes such surveys annually. However, if there continues to be a great proliferation of commercial and government surveys in a community, households might begin to be called on more frequently, and local governments might begin to have a problem in their own surveys.

Of a more general concern is the feeling occasionally expressed that surveys will sensitize citizens to problems about which they had not previously been concerned. That is, by asking citizens about such issues as the cleanliness of their streets, the frequency of appearance of rats in their neighborhoods, or the adequacy of street lighting, they may become more critical of government services. Though only a small proportion of the population is likely to be surveyed in any one survey, the findings of the survey may be widely publicized. (The information, in general, is public information and is often picked up by the local newspapers.) Indeed it does seem possible that citizens will become more sensitive to the issues included in the survey and publicized. Public officials may feel that the citizen survey is likely to raise the heat level and cause them more problems than the many they already have. Such a view cannot be dismissed lightly. However, the authors suspect that the great majority of the population would look upon this as a weak and even contrary excuse for not obtaining citizen feedback.

A related concern is that the results would generate substantial amounts of adverse publicity; that the press or political opponents would highlight inadequacies or pull findings out of context. This, indeed, seems bound to happen occasionally. The authors, in their experiences, have been pleasantly surprised to find that reporters for the various information media generally acted quite responsively in reporting citizen survey results. But nevertheless, the danger is there.

A FEW SUGGESTIONS TO LOCAL GOVERNMENTS
Some recommendations for local governments' consideration in conducting surveys are:

(1) Ask questions on which citizens are competent or reasonably able to answer.

(2) Include questions on conditions or aspects of service that can be affected by the government or else which provide information essential in interpreting survey findings or classifying respondents.

(3) For the purpose of regularly assessing the quality of services and obtaining planning data, regular (perhaps annual) surveys should include similarly worded questions so that changes and trends over time can be identified.

(4) Use professional survey help (for example, there are numerous commercial organizations and universities that have such capabilities) to help develop the questionnaire, to select the sample, to conduct the interviews (or at least to train and supervise volunteer or government employee interviewers) and to help analyze the data.

The complete survey task is likely to be beyond the technical capabilities of most local governments.

(5) Provide for the analysis and use of the data. A survey will collect a great deal of raw data which needs time consuming analysis to make it most useful.

CONCLUDING JUDGMENT

Despite the many technical problems, and the possibility for manipulation and copout, the citizen survey properly used appears to be a potentially major tool for local (and state) executive and legislative officials for planning, assessing, and revising programs. If one believes that government services are primarily for the purpose of providing services and assistance to the citizens of the community, direct and representative feedback through the use of systematic surveys of samples of citizens would seem to be one major way for governments to assess its own performance and obtain meaningful information on citizen viewpoints.

NOTES

---

1. For advantages and special considerations of citizen surveys for local governments, the reader is referred to a report by Kenneth Webb and Harry P. Hatry, Obtaining Citizen Feedback: The Application of Citizen Surveys to Local Governments (Washington, D.C.: The Urban Institute 1973).

2. L. R. O'Neall and A. Gollin, City Government and the Quality of Life in Richmond: Findings of a Citizen Survey (Washington, D.C.: Bureau of Social Science Research, Inc. 1973).

3. The basic questionnaire used in St Petersburg and Nashville which contains many of the questions used in Palo Alto, Dallas and Randolph Township is contained in a report published by The Urban Institute and the International City Management Association entitled Measuring The Effectiveness of Basic Municipal Services: Initial Report (Washington, D.C. 1974).

4. The findings from the St Petersburg surveys of 1973 and 1974 have been published by St Petersburg, Florida's Office of Management and Budget, Multi Service Citizen Survey for the City of St Petersburg, 1973, and 1974 Multi-Service Citizen Survey for the City of St Petersburg. The results of selected findings from the Nashville 1973 survey were incorporated, along with other

sources of effectiveness measuring data into a report to its citizens, *How Well is Metro Doing?* Metropolitan Government of Nashville-Davidson County, Department of Finance (1974). See also: Township of Randolph (N.J.), Report on The Survey of Citizen Attitudes (August 1975).

　　5. 'A Primer on Poll-Taking', *Newsweek* (5 January 1976), 16-17.

　　6. See, for example, 'How to Save Urban America, Regional Plan Association Choices for '76', The New American Library (1973).,

Notes on Contributors

Louis H. Blair is on the senior staff of the State and Local Government Research Program of the Urban Institute in Washington, D.C.

Terry Nichols Clark is Associate Professor of Sociology at the University of Chicago, and has taught at Columbia, Harvard, Yale and the Sorbonne. He has been a consultant to the Urban Institute and the Department of Housing and Urban Development. He is the author or co-author of *Community Structure and Decision Making* (1968), *Community Politics* (1971), *Community Power and Policy Outputs* (1973), and *Comparative Community Politics* (1974).

G. David Curry is a doctoral candidate in sociology at the University of Chicago. He is pursuing work on roll call analysis of the US Congress and the American military.

Harry P. Hatry is Director of the State and Local Government Research Program of the Urban Institute in Washington, D.C. He has written extensively on local government productivity, performance measurement, citizen surveys, and related issues. His paperback, *Obtaining Citizen Feedback: The Application of Citizen Surveys to Local Governments* (1973), with Kenneth Webb, is one of the most widely used works on the subject by local officials in the US.

Wayne Hoffman holds a PhD in political science from the University of Chicago and is currently Assistant Professor at the School of Government and Public Administration of The American University.

John P. McIver is a research associate of the Workshop in Political Theory and Policy Analysis, Indiana University.

Elinor Ostrom is a Professor in the Department of Political Science and the Workshop in Political Theory and Policy Analysis at Indiana University. She is co-author of *Community Organization and the Provision of Police Services* and author of articles published in the *Urban Affairs Quarterly*, *Social Science Quarterly*, *Public Choice*, *American Journal of Political Science*, *Public Administration Review*, and *Journal of Criminal Justice*.

Jerome Rothenberg is Professor of Economics at the Massachusetts Institute of Technology and previously taught at Northwestern University and the University of Chicago. His theoretical works on collective decision-making include *The Measurement of Social Welfare* (1961); more applied works include *Economic Evaluation of Urban Renewal* (1967). He has published important articles on residential choice decisions and related issues in urban economics.

Douglas Scott is a political scientist on the staff of the Rand Corporation. He previously held positions at the University of Michigan and the University of California, Los Angeles.

T. Nicolaus Tideman is Associate Professor of Economics at the Virginia Polytechnic Institute and State University. He previously taught at Harvard. He has completed several important papers on preference revelation for public goods as well as work on urban economics.

NOTES

NOTES